U0612935

设施火龙果栽培技术
Protected Cultivation of Dragon Fruit

（中英双语）
（Chinese-English Bilingual）

王国章　主编
Chief Editor Wang Guozhang

中国农业出版社
China Agriculture Press
北　京
Beijing

图书在版编目（CIP）数据

设施火龙果栽培技术：汉、英／王国章主编．——
北京：中国农业出版社，2024.4
ISBN 978-7-109-31898-4

Ⅰ.①设… Ⅱ.①王… Ⅲ.①热带及亚热带果—果树
园艺—设施农业—汉、英 Ⅳ.①S667.9

中国国家版本馆 CIP 数据核字（2024）第 076237 号

设施火龙果栽培技术（中英双语）
SHESHI HUOLONGGUO ZAIPEI JISHU（ZHONGYING SHUANGYU）

中国农业出版社出版
地址：北京市朝阳区麦子店街 18 号楼
邮编：100125
责任编辑：李 瑜 黄 宇
版式设计：王 晨 责任校对：吴丽婷
印刷：中农印务有限公司
版次：2024 年 4 月第 1 版
印次：2024 年 4 月北京第 1 次印刷
发行：新华书店北京发行所
开本：880mm×1230mm 1/32
印张：4.25
字数：118 千字
定价：48.00 元

版权所有·侵权必究
凡购买本社图书，如有印装质量问题，我社负责调换。
服务电话：010-59195115 010-59194918

编 译 人 员
Editors and Translators

主　编　王国章

副主编　崔凯阳　郑晓红

翻　译　罗欣蓓　陈　虹

审　稿　潘芝梅　[德] 丽萨·妮迟　张　泠

Chief Editor　　　　Wang Guozhang

Deputy Editor　　　Cui Kaiyang

　　　　　　　　　　　Zheng Xiaohong

Translator　　　　　Luo Xinbei

　　　　　　　　　　　Chen Hong

Riviser of the Translation

　　　　　　　　　　　Pan Zhimei

　　　　　　　　　　　Lisni Schutzfolie Abziehen

　　　　　　　　　　　Zhang Ling

前言

　　火龙果作为一种有着独特的观赏、营养和经济价值的热带水果，深受人们喜爱，不仅在热带地区广泛栽培，而且在许多温带地区设施栽培。浙江省余姚市第二职业技术学校师生已经实施火龙果栽培试验近 7 年，获得较好效益。为了助力产业振兴，促进区域农民共同富裕，助力"一带一路"建设，更好指导职高学生和农民生产与实践，我们编写了这本产教融合的专业读物。

　　本书采用图文并茂的形式，介绍了火龙果生物学特性，以及我国主栽的火龙果三大类、40 个品种，对火龙果普通栽培技术，如繁殖技术、建园、幼树管理、整形修剪、人工授粉、病虫害防治、采收，以及火龙果现代化栽培技术均做了详细的介绍。

　　本书由浙江省特级教师王国章主编和统稿，王国章编写了第一章火龙果生物学特性和第二章火龙果品种，崔凯阳编写了第三章火龙果普通栽培技术，郑晓红编写了第四章火龙果现代化栽培技术。

　　本书作为"一带一路"建设项目图书，适合中等职业学校涉农专业学生选学和农民培训，也可供火龙果爱好者参考。

　　本书编写过程得到宁波市教育局相关领导的指导，宁波市职业教育与成人教育学院吕冲定书记，余姚市第二职业技术学校陈建栋校长、陈少平副校长、刘庆国主任均给予了大力支持，宁波市四明职业高级中学马赤骏老师、周海光老师，宁波市北仑职业高级中学王佳翘老师，慈溪技师学院戴利民均提供了大量资料，丽水职业技术学院潘芝梅教授等人对本书进行审稿，在此向各位领导、专家表示衷心的感谢。

由于编写时间紧迫，加之编者水平有限，不足之处在所难免，望读者朋友批评指正。

编　者

2023 年 5 月

FOREWORD

As a tropical fruit with unique ornamental, nutritional and economic value, the dragon fruit is deeply appreciated by people. It is not only widely cultivated in tropical regions, but also cultivated in many temperate regions with special facilities. The teachers and students from Yuyao No. 2 Vocational Technical School in Zhejiang Province have been engaged in the dragon fruit cultivation experiments for nearly seven years and have achieved fruitful results. To empower the revitalization of the industries, promote the common prosperity of farmers in different regions, contribute to the "Belt and Road" Initiative, and better guide the production and practice of vocational high school students and farmers, this professional book integrating production and education has been compiled.

This book adopts both pictures and texts to introduce 40 varieties of dragon fruit in three major categories cultivated in China, including detailed explanation about common cultivation techniques such as propagation technology, establishment of farms, sapling management, shaping and pruning, artificial pollination, disease and pest prevention and control, harvesting, as well as modern cultivation techniques.

This book is edited and drafted by Wang Guozhang, a special-grade teacher of Zhejiang Province. Wang Guozhang writes the introduction part of biological characteristics and varieties of the dragon fruit; Cui Kaiyang writes the part of the common cultivation technology of dragon fruit; Zheng Xiaohong writes the part of the modern cultivation technology of dragon fruit.

As a "Belt and Road" project book, this book is for the secondary vocational school students majoring in agriculture and the training of farmers, and can also serve as a reference for those interested in the dragon fruit.

The compilation of this book has gained the great support from leaders from the Ningbo Municipal Bureau of Education, Lv Chongding, the secretary of Ningbo Vocational and Adult Education College, Chen Jiandong, the principal of Yuyao No. 2 Vocational and Technical School, Chen Shaoping, the vice principal of Yuyao No. 2 Vocational and Technical School, Liu Qingguo, the director of Yuyao No. 2 Vocational and Technical School. Besides that, Ma Chijun and Zhou Haiguang from Ningbo Siming Vocational High School, Wang Jiaqiao from Ningbo Beilun Vocation High Middle School, and Dai Limin from Cixi Hangzhou Bay Secondary Vocational School provide exhaustive materials for this book. Professor Pan Zhimei from Lishui Vocational and Technical College and others review the draft. Our sincere gratitude goes to all the above-mentioned leaders and experts.

Due to the tight schedule and the limited knowledge, this book has yet to be perfected, we welcome suggestions and critiques from the readers.

Editors

May, 2023

目录

 CONTENTS

Foreword

第一章
火龙果生物学特性

第一节　火龙果概述

一、火龙果的科属

火龙果，又名红龙果、仙蜜果，仙人掌科多年生量天尺属攀缘性肉质植物。火龙果原产中南美洲，主要分布在中美洲至南美洲北部，世界各地广泛栽培。

二、火龙果的营养价值

火龙果外观独特，呈椭圆形，外观为红色或黄色，色泽艳丽，营养丰富，风味独特，是一种低热量、高纤维、高营养、清热下火的水果，有着"美容皇后"的美誉，深受消费者喜欢。

火龙果花不但富含氨基酸、微量元素等，而且含有大量的功能性成分，功效独特。

火龙果营养丰富，味甜多汁，含有丰富的维生素、葡萄糖及人体所需钾、钙、镁等矿物质，它还含有一般植物少有的植物性白蛋白，以及花青素、维生素C和水溶性膳食纤维。白蛋白是具黏性、胶质性的物质，对机体重金属中毒具有解毒功效，对胃壁还有保护作用；花青素在火龙果中的含量较高，尤其在红肉种类的果实中，它具有抗氧化、抗自由基、抗衰老的作用，还能预防机体脑细胞变性，有助于抑制痴呆症的发生；维生素C具有美白皮肤的作用；丰富的水溶性膳食纤维具有减重、降低血糖、润肠、辅助预防大肠癌等作用。

火龙果籽油中的不饱和脂肪酸含量高达 78.10%，其中单不饱和脂肪酸含量 33.39%，多不饱和脂肪酸含量 44.71%；亚油酸含量高达 44.29%，高于菜籽油和花生油，和芝麻油相近；油酸（单烯类不饱和脂肪酸）含量高达 31.75%。

火龙果果皮、果肉中含有大量红色素，是提取加工天然食用红色素的良好来源。火龙果果皮红色素具有水溶性和醇溶性，在酸性条件下稳定，pH 1.5 时呈现鲜艳的红色，可溶于乙醇、丙酮、异丙醇、乙酸、柠檬酸、酒石酸和 0.2 摩尔/升的 HCl 溶液，微溶于乙醚和乙酸乙酯，具有极性或弱极性分子物质特性。

三、火龙果的经济价值

火龙果是一种集食用、观赏为一体的热带水果，既可以盆栽，又可以大田种植。火龙果一年四季均可种植，具有多批次开花结果习性，产量高。

台湾是我国最早进行火龙果品种引进的地区，而广东是大陆地区最早种植火龙果的省份。

目前，国内火龙果主要有广东、广西、海南、云南、贵州五大产区，广西南宁火龙果种植面积 15 万亩*，是当之无愧的全国最大火龙果生产基地，福建、辽宁、四川、上海、浙江等地也有种植。

第二节　火龙果的生物学特性及生长特点

一、火龙果生物学特性

1. 分类

火龙果（*Hylocereus undulatus* Britt），属双子叶植物纲、仙人掌目、仙人掌科、柱状仙人掌亚科、量天尺属（图 1-1）。

---

* 亩为非法定计量单位，1 亩＝1/15 公顷。余后同。——编者注

图1-1　火龙果

目前，水果市场上出现一些被人们统称为火龙果的红龙果、紫龙果和金龙果。红龙果和紫龙果的果实外被为红色，果肉分别为红色和紫色，金龙果的果实外被为黄色，果肉为白色。尽管它们被统称为火龙果，但它们却是同属不同种的植物，红龙果种名是 *H. polyrhizus*，紫龙果种名是 *H. costaricensis*，金龙果种名是 *H. megalanthus*。本属植物约有 20 种，生长在中美洲和墨西哥等地。下面介绍一些主要品种的形态特性和繁殖栽培方法。

火龙果为攀缘植物，有附生习性，能自花传粉而结实。喜温暖、湿润环境。生长适温 25～35 ℃。对低温敏感，在 3 ℃以下，茎节容易受病腐烂。喜含腐殖质较多的肥沃壤土，盆栽用土可用等量的腐叶土、粗沙及腐熟厩肥配制。火龙果多用扦插繁殖，操作时，可在生长季节剪取生长充实或较老的茎节，切取不短于 15 厘米的切段，切后晾几天，待切口干燥后插于沙床或直接插于土中，1 个月后能生根，根长 3～4 厘米时可移栽到小盆或直接于露地栽培。生长季节需充分浇水，每半个月追施腐熟液肥 1 次。冬季应节制浇水并停止施肥。可盆栽用于观赏，做水果栽培时宜采用大田栽培。

火龙果为热带植物，在温暖湿润、光线充足的环境下生长迅速，能耐 0 ℃低温和 40 ℃高温，为保证其常年生长和多次结果，应使生长环境尽量达到适宜温度 25～35 ℃。春夏季露地栽培时应多浇水，使其根系保持旺盛生长状态，在阴雨连绵天气应及时排水，以免感染病菌造成茎肉腐烂。火龙果也适应多种土壤，但以含腐殖质多、保水保肥的中性土壤和弱酸性土壤为好，为使其种植后生长旺盛，多施发酵过的有机肥作基肥，苗期多施钙镁磷肥和复合肥，开花结果期要增补钾、镁肥，以促进果实糖分积累，提高品质。火龙果栽后 12～14 个月开始开花结果。栽植后第二年每柱产果 20 个以上，第三年进入盛果期。火龙果在栽培过程中，要注意摘心，促进分枝，并让枝条自然下垂，以利多结果且易于采收。

红龙果，多分枝，长可达 2 米，茎 3 棱，有时 2 棱或 4 棱。暗绿色，棱边缘有刺座，有 2～6 根短刺。花大型，白色，直径 25～30 厘米，晚间开花。自然状态下较难结实，采用火龙果的花粉授粉时能使果实增多，但果肉保持红色。红龙果的市场价格比火龙果高，但产量略低。

紫龙果，茎 3 棱，棱边缘有刺座，有 2～4 根短刺。花白色，有香气，直径 30 厘米左右。自然状态下也较难结实，采用火龙果的花粉授粉时能使果实增多，但果肉保持紫红色。国内已有少量紫龙果栽培。

金龙果，可能是仙人掌科最大花的植物。植株附生，攀缘匍匐生长，茎长可达 3 米，3 棱，棱边轻微卷曲，白色刺座，有 1～3 根短刺。花白色，有香气，直径 32～38 厘米。果被黄色，果肉白色，口感好，国内水果市场上已有少量进口，还未有栽培。

2. 花

火龙果的花为雌雄同体，一般一花一果。花蕾似漏斗，花萼黄绿色，肉质厚，呈鳞片状，花多在晚上开放，花瓣纯白色，少数种类为红色，雄蕊细长且多，花粉乳黄色，雌蕊柱头青色。含苞待放的花蕾长达 30 毫米以上，花瓣纯白色，花蕾黄绿色，光洁硕大，花长 30 厘米，有"大王花"之美称（图 1-2）。

图 1-2 火龙果的花

3. 果实

火龙果果实形态奇特，极具观赏价值：果实近球形，体态雅丽，在生长时，颜色由绿慢慢变红（除黄皮种外），成熟时颜色鲜红或玫瑰红。单果重 350～500 克，多数红色，许多种类可食用。花芽分化至开花时间相对较短，一般为 40～50 天。自花授粉，但坐果率低，一般可用人工授粉，用手指蘸取花粉涂于柱头即可。授粉后 25～30 天，果皮颜色转红，果皮转红 4～6 天即可采收。植株寿命长，可达 100 年以上，一定年限内，产量随树龄增长而增加。

4. 根

火龙果为浅根系的肉质植物，种植时不需挖果坑，种入深约 3 厘米，浇水保湿，拌匀基肥，立上攀缘柱，即可种植，成活率可达 100%。

5. 茎

本属植物的节状茎，通常只有 3 棱，主茎一般可产生 8 条主要分枝，直径可达 13～18 厘米，茎节有攀缘根，每段茎节凹陷处有短刺 1～3 根。匍匐或攀缘生长，长可至 10 米以上，茎部气根多。刺座稀，刺锥状，很短。茎 2～3 棱，叶深绿色，棱边呈波浪状有小刺。枝叶融为一体，呈棱剑状。

6. 生长环境

火龙果为热带水果，喜光耐阴、耐热耐旱、喜肥耐瘠。在温暖湿润、光线充足的环境下生长迅速，春夏季露地栽培时应多浇水，使其根系保持旺盛生长状态，在阴雨连绵天气应及时排水，以免感染病菌造成茎肉腐烂。其茎贴在岩石上亦可生长，植株抗风力极强，只要支架牢固可抗台风。火龙果耐 0 ℃低温和 40 ℃高温，生长的最适温度为 25～35 ℃，冬季气温低于 8 ℃的地区不宜露地栽培。火龙果可适应多种土壤，但以含腐殖质多、保水保肥、排水性好的中性土壤和弱酸性土壤为好。

二、火龙果的生长特点

1. 适应性强

火龙果植株的外形与霸王花（野生三角柱）极为相似，只要霸王花能生长的地方，均可种植火龙果。平地、山坡地、水田、旱地均可种植。仙人掌科植物是热带植物，因此在我国只适宜在北回归线以南地区种植。长时间的霜冻对植物生长有一定的影响。

2. 病虫害少

火龙果是一种较少发生病虫害的果树。在幼苗期，因施用农家肥而引来大蚂蚁对幼苗咬食，用灭蚁药除去即可；当果树长大后，尤其是从开花至果实成熟，均无需喷洒任何农药。在种植地，几乎所有的水果果实都要套袋，唯有火龙果不需套袋。

3. 高产性好

火龙果的栽培技术较为简单，3 年以上的果园产量可达 825 千克/亩以上。从果园试种的情况看，要达到这样的产量并不难，需亩栽110 株，每株结果 7.5 千克。一般单果重在 0.25 千克以上，每株有 40 条以上的结果枝即可。每个结果枝有数个花蕾，需要人工摘去多余的花蕾。

4. 产果期长

火龙果高产的同时，产果期也长达 180 天，可分批成熟，这样便可保证水果均衡上市，使果农容易获得稳定的收入。从试种情况看，从开花至采果约 40 天，即使果实熟透也不会从枝上自然脱落。

5. 果实耐贮运

火龙果采收后，在常温下可贮放 14～21 天，而冷藏保鲜期可达数月之久。由于果皮厚并有蜡质保护，运输中不易被碰坏，耐贮运。在保质期内，果皮的红色会变得越来越鲜艳。

 第二章

火龙果品种

火龙果有 3 个亚种：红肉亚种、白肉亚种、黄肉亚种。目前我国主栽红皮白肉、红皮红肉、黄皮白肉火龙果 3 种类型，还有青皮红心、双色火龙果等。

第一节 红龙果

一、红龙果总述

红龙果果肉红色或紫红色，鲜食品质佳（图 2-1）。在台湾、广东、广西产期为 5—11 月，海南产期更长。红肉火龙果自花授粉坐果率低，需要白肉类火龙果授粉，坐果率高且果大。

果实可加工成果汁、果粉、红色色素、冰激凌粉、果冻和果酱。

红肉火龙果卵形、圆形、圆筒形，果皮鲜红色、

图 2-1 红龙果

有光泽，果肉红色或紫红色，细腻而多汁，含可溶性固形物 16%～21%，甜度较高。含有大量花青素，花青素是一种抗氧化剂，抗氧化能力比胡萝卜素强 10 倍以上。单果重 350～1 000 克，亩产 1 000～3 000 千克。

二、红龙果代表品种

1. 粉红龙

粉红肉类火龙果（图2-2）。由贵州省果树科学研究所从火龙果新红龙发现的芽变单株选育而成。果实椭圆形，果皮红色，厚度0.29厘米，果肉粉红色，含甜菜红素，种子黑色，平均单果重340克，亩产1834.7千克。适宜在贵州省年平均温度18.5℃以上，常年1月气温高于−1℃的区域种植。

图2-2　粉红龙

2. 金钻

红皮红肉类火龙果，又名女王头（图2-3）。自花授粉，台湾品种，单果重800克以上，最大可至2 500克，亩产4 000千克以上。可溶性固形物含量17%～20%。4月始现花，每批花间隔天数为15天左右，花期至10月底结束，共18批花左右。采果期6月下旬至12月上旬，每年可采8批次，开

图2-3　金　钻

花至采收需30～40天。适合在广西、广东、福建、海南、云南种植。

3. 祥龙

红皮红肉类火龙果（图2-4），自花授粉，台湾品种。平均单果重418.6克，最大单果重超过1 000克。可溶性固形物含量

16.5%。扦插苗第四年进入盛果期，株产8千克以上。4月中旬现蕾，花期从4月下旬延续至11月上中旬。采果期6月下旬至12月上旬，每年可采果8批次，火龙果从开花至采收需35～40天。适合在广西、广东、福建、海南、云南种植。

图2-4　祥　龙

4. 喜香红

红肉类火龙果（图2-5），自花授粉，台湾品种，单果重500克，株产6.2千克。果圆形，果皮鲜红，果肉血红、花青素含量高、细嫩、高糖超甜、味甜多汁略带花香，可溶性固形物含量18%～22%，适合在采摘园发展，果皮厚，耐贮存和运输。自4月中旬出花蕾直到12月初最后一批

图2-5　喜香红

花谢。一般幼苗定植10～12个月，枝干超过1米就可以开花结果。每年的5—11月都是产果期。适合在广西、广东、福建、海南、云南种植。

5. 台湾大红

红皮红肉类火龙果（图2-6）。自花授粉，台湾品种。单果重500克，最大可达1500克。亩产4000千克以上，可溶性固形物含量18%～22%。自4月中旬出花蕾直到12月初最后一批花谢，一共开16批花左右。果实夏季在树上全果15天以上而不会裂

果，冬季可坐果 2 个月，适合在广西、广东、福建、海南、云南种植。

6. 石火泉

红皮红肉类火龙果（图 2-7）。自花授粉，台湾品种。平均单果重 446 克，最大可达 1 050 克。第四年株产 7.8 千克以上，可溶性固形物含量 20%。5 月上中旬现蕾，花期从 5 月下旬持续至 10 月中旬。果实成熟期从 7 月上旬至 11 月下旬，全年单株可采果 4～8 批次，果实生长发育期 30～40 天。适合在广西、广东、福建、海南、云南种植。

图 2-6 台湾大红　　　　　图 2-7 石火泉

7. 贵紫红龙

红皮红肉类火龙果（图 2-8）。自花授粉，由紫红龙芽变而来。单果重 215 克，亩产 870 千克，可溶性固形物含量 13.8%。5 月上旬现蕾，花期从 5 月中旬持续至 10 月中旬。每年开花结果 8～10 批次，第一

图 2-8 贵紫红龙

批花 5 月上旬现蕾，5 月中下旬开花，6 月下旬果实成熟，最后一

批花 9 月下旬现蕾，10 月中旬开花，11 月下旬至 12 月上旬成熟。适宜贵州常年 1 月最低温度 0 ℃以上的区域种植。

8. 软枝大红

红皮红肉类火龙果（图 2-9）。自花授粉。单果重 500～1 000 克，亩产 4 000～5 000 千克，可溶性固形物含量可达 20%以上。5 月上旬现蕾，花期从 5 月中旬持续至 10 月中旬。全年坐果 13～15 批，6 月至

图 2-9 软枝大红

翌年 2 月都有果收。适合在广西、广东、福建、海南、云南种植。

9. 蜜红

超大果型红肉火龙果（图 2-10）。我国台湾选育。果实长圆形，深紫红色，种子黑芝麻状，口感好，贮藏保鲜期较长。单果重 650 克，最大可达 1 540 克，盛果期，株产 6.6 千克，无大小年。3 月上旬始发春梢，进入结果期后新梢停止萌发；

图 2-10 蜜 红

具有多次开花结果习性，5 月上旬现蕾，花期从 5 月中旬持续至 10 月中旬，果实成熟期从 7 月上旬至翌年 1 月中旬，单株全年可采 6～12 批次，果实生长发育期 25～40 天。适宜海南、福建漳浦种植。

10. 黔果 1 号

紫红龙的大果型紫红肉芽变新品种（图 2-11）。果实椭圆形，果肉紫红色，种子黑色，单果重 460 克，最大单果重 786 克，可溶性固形物含量 13.6%，果实着色好，不易裂果，风味浓。每年开

花结果 8～10 批次，第一批次 5 月上旬现蕾，5 月中下旬开花，6 月下旬成熟，最后一批 9 月下旬现蕾，10 月中旬开花，11 月下旬至 12 月上旬成熟。适合在贵州常年 1 月最低温度 0 ℃以上的区域种植。

图 2-11 黔果 1 号

11. 富贵红

红肉类火龙果（图 2-12）。自花授粉率达 100%，我国台湾选育的自交亲和型优良新品种。果大，椭圆形，果皮较薄，单果重 445.6 克，最大可达 1 000 克以上，第四年株产 7.8 千克以上，成熟果的果皮色泽艳丽，呈玫瑰红色，果肉紫红色，肉质软脆，汁多，果实可溶性固形物含量 21%，适应性强、无大小年。3 月下旬至 5 月中旬萌发春梢，进入结果期后，新梢才停止萌发；5 月上中旬现蕾，花期从 5 月下旬持续至 10 月中旬，果实成熟期为 7

图 2-12 富贵红

月上旬至 11 月下旬，产量主要集中在 7—8 月，全年单株可采 6～12 批次果，果实生长发育期 30～40 天。适宜在广西、广东种植。

12. 蜜玄龙

台湾新品种，与台湾大红同为姐妹品系（图 2-13），果形偏圆，自花亲和、花内授粉，夏果大果率高，冬果单果重最大可达 1.8 千克，不易裂果，肉色红，果实有香气，汁多、籽少、甜度高，肉质清爽带脆，品感佳。

图 2-13 蜜玄龙

13. 大丘 4 号

红肉类火龙果（图 2-14）。需人工授粉。由广州大丘有机农产有限公司、广东省农业科学院果树研究所培育。果实近圆球形、单果重 316 克，可溶性固形物含量 9.0%，五年生果园亩产 2 452 千克。

图 2-14 大丘 4 号

适宜在广东省中部及南部地区种植。

14. 红绣球

果实圆形（图 2-15），单果重 540 克，最大果重 1 350 克。成熟果实鲜红色。果皮厚 0.3 厘米，易剥离。果肉红色，细腻，果肉中有黑芝麻状细小种子，汁液多，可溶性固形物含量 24%。3 月定植苗木，10～12 个月后就可以开花结果，每年开花

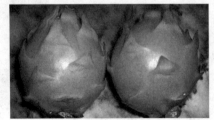

图 2-15 红绣球

结果12～15次，第三年亩产2 947千克。温室栽培一年四季均可生长。

15. 美龙2号

红翠龙的芽变后代，自花授粉，结果率90％（图2-16）。果实 近球形，皮红色带紫，皮厚不裂果，果肉深紫红色 ，肉质细腻，易流汁，可溶性固形物含量20％，单果重500～1 000克，亩产1 475 千克。1年可采

图2-16 美龙2号

12～14批次果，其中大批次果约6批。头批果6月中旬成熟，末批果于12 月上旬成熟，夏季花后30～35天果实成熟。成熟果留树期10～30 天。适合在广西南宁露地栽培。

16. 红金宝

自花授粉，台湾特选品种（图2-17），果皮厚，不裂果，耐贮运。鳞片组成独特，外观像红宝石火焰，单果重1 300克以上，最大单果重2 200克，为目前红龙果内"最巨大的王者"。果肉带有轻微香气，可溶性固 形物含量23％，生产时间为每年6 12月，如通过技 术推迟花期，果实可延长至翌年2月采收。

图2-17 红金宝

适合在台湾和广西的南宁、百色地区种植。

17. 吉人玫瑰

由南宁市大吉利农业有限责任公司从以色列优质火龙果与本土

野生优质火龙果培育而成（图 2 - 18）。果肉清脆甜美，有玫瑰香味。可溶性固形物含量 22% 以上，花青素含量高，在树上成熟后再采摘，果实鳞片已很红、鳞片薄而皱，采摘后不再进行任何化学保鲜处理。

18. 桂香红

自花授粉，台湾品种（图 2 - 19），单果重 500 克左右，亩产 4 000 千克，可溶性固形物含量 22%。香甜可口，甜而不腻，不易裂果，耐储存。栽后 12～14 个月开始开花结果，每年可开花 12～15 次，火龙果从开花至果实成熟，约 30 天。上市时间为 6—12 月，抗病性较强，几乎不使用任何农药，可作有机绿色水果。适合在广东、广西、云南、贵州大部分地区种植。

图 2 - 18　吉人玫瑰　　　　图 2 - 19　桂香红

19. 香蜜龙

自花授粉，雄雌花柱同高，坐果率高，最大果重 1 000 克（图 2 - 20）。亩产 5 000 千克。花青素含量高，果肉偏紫色，细腻，有玫瑰花香，可溶性固形物含量可达 23%，在 5 月

图 2 - 20　香蜜龙

下旬至 6 月上旬开花，持续至当年 11 月中旬，其间可收果 8～13 批，从开花当日到果实成熟采收，需 30～40 天。抗病性好，既能适应南方露地种植，也可以在北方大棚种植。

20. 桂红龙 1 号

红肉类火龙果（图 2-21）。自花授粉，由广西壮族自治区农业科学院园艺研究所用普通红肉火龙果的芽变单株选育而来。果皮玫瑰红色，果肉深紫红色，肉质细腻，易流汁、味清甜，略有玫瑰香味，可溶性固形物

图 2-21 桂龙红 1 号

含量 22%。单果重 533.3 克，四年生果树亩产 2 869.75 千克。果实成熟后留树期达 15 天以上，抗逆性良好，可做抗寒选育，遗传性状稳定，可单一品种规模种植。适合在广东、广西、海南种植。

21. 美龙 1 号

红肉类火龙果（图 2-22）。自然授粉结果率 92%。为广西壮族自治区农业科学院园艺研究所、南宁振企农业科技开发有限公司从哥斯达黎加红肉和越南白玉龙杂交组合后代实生苗中筛选出的优良单株。果实长圆形，皮鲜

图 2-22 美龙 1 号

红，果肉细腻，清甜脆口微香。果实属中大型，单果重 525 克，第三年果树亩产 1 870 千克。头批果 6 月中旬成熟，末批果于 12 月底成熟，夏季花后 30～35 天成熟。适合在广西、广东种植。

22. 桂热 1 号

红肉类火龙果（图 2-23）。自花授粉，是广西壮族自治区亚

热带作物研究所、广西壮族自治区山区综合技术开发中心从桂红龙1号选取的变异株。果皮鲜红色，果肉紫红色，苞片不带刺。单果重600克以上，第三年亩产2015千克。适合在广西种植。

图 2-23 桂热 1 号

23. 嫦娥 1 号

红肉类火龙果（图 2-24），自然授粉。果实近长圆形，果皮玫瑰红色，果肉深红色，肉质细腻，汁多，味清甜。单果重 410克，亩产量 2865 千克。适合在广西种植。

24. 红冠 1 号

红肉类火龙果（图 2-25）。由华南农业大学园艺学院、东莞市林业科学研究所从广州市从化区大丘园农场引进的红水晶火龙果实生繁殖群体中单株优选而成。果实近球形，平均单果重 308 克，果皮鲜红色，果肉紫红色，肉质软滑，味清甜，可溶性固形物含量10.1%。第三年亩产 1628 千克，适宜在广东中部及南部地区种植。

图 2-24 嫦娥 1 号

图 2-25 红冠 1 号

25. 紫红龙

紫红肉类火龙果（图2-26）。贵州省果树科学研究所从新红龙火龙果中发现的芽变单株，经系统选育而成。果实圆形，果肉紫红色，种子黑色，平均单果重330克，第三年亩产2000千克。四季均能生长，每年结果10~12批

图2-26　紫红龙

次，从现蕾到开花15~21天，从开花到果实成熟28~34天。适宜在贵州年均温度18.5℃以上，常年1月气温高于-1℃的区域种植。

26. 紫龙

红皮红肉类火龙果（图2-27）。自花授粉，由台湾种植户从东南亚、中美洲等地引进及从本地收集的数十份红肉火龙果种质资源中选育。单果重245~850克，第三年亩产2700千克。可溶性固形物含量15%。

图2-27　紫　龙

花果期为3月下旬至11月中旬，盛产期每年开花结果12~15批次，单次开花结果周期45天，存在明显的花果重叠特点，同时期最多有3批花果。高温期花后35天成熟；低温期花后45天成熟；一年有相对集中的花期4次，可以分为16~20次开花，一枝一次可以开花3~5朵。果实生长发育期35~40天。适合在海南、广西、广东、福建、云南种植。

27. 金都1号

红皮红肉类火龙果（图2-28）。自花授粉，由广西南宁金之

都农业发展有限公司从中南美洲火龙果原种与红肉种火龙果的杂交后代中选育而成。果实短椭圆形，果皮紫红色，果肉深紫红色，肉质细腻，味清甜，有玫瑰香味。单果重575克。可溶性固形物含量22％。亩产3 820千克。4

图2-28　金都1号

月下旬现蕾，花期从5月中旬持续至10月下旬。果实成熟期为6月下旬至12月下旬，单株全年可采8～12批次果，果实生长发育期35～50天。适合在广西桂南、桂东南地区以及百色河谷地区种植。

28. 台农3号

自花授粉。果实圆整，叶状鳞片为红色（图2-29），果实皮薄，果肉为红色、柔绵细嫩，单果重490克，在鲜食火龙果品种中品质最佳。能耐0 ℃低温和40 ℃高温，生长的最适温度为25～35 ℃。在我国的海南、广西、广东、福建、云南等地均可种植。

图2-29　台农3号

29. 莞华红

红肉类火龙果（图2-30）。东莞市林业科学研究所、华南农业大学园艺学院从红水晶火龙果实生繁殖群体中通过单株优选而

图2-30　莞华红

成。果实近椭圆形至球形，果皮鲜红色，单果重448克，果肉紫红色，果皮厚0.2厘米。肉质细腻软滑、风味浓郁，不易裂果，耐贮运。三年生果树亩产1 737千克。适宜在广东中部及南部地区种植。

第二节　白龙果

一、白龙果总述

白龙果红皮白肉，自花授粉，结果率高，抗病力强。果实表面无刺，果皮厚，果肉比较清甜，耐贮运。鲜食品质一般，可加工成果汁、果粉及果酱。

白龙果包括：野生火龙果、普通白玉龙、新白玉龙、红宝石和黔白1号等，以越南1号品种最好。果实成熟期在6—10月，适合在台湾、广东、广西种植。

二、白龙果代表品种

1. 白玉龙

红皮白肉类火龙果（图2-31）。自花授粉，台湾品种。果皮紫红色，厚0.2厘米；果肉白色，肉质清脆、多汁，可溶性固形物含量15%。单果重425克，最大果重超过1 000克，第四年进入盛产，株产7千克

图2-31　白玉龙

以上。3月下旬至5月下旬大量冒芽，5月下旬现蕾，从现蕾到开花13~18天，花期从6月中旬延续至9月下旬，花期持续4个多月；盛花期集中在6月下旬至7月上旬、8月上旬至9月下旬；采果期在7月下旬至11月中旬，每年可采6批次果，从开花至果实成熟的生育期为35~40天。适合在广西、广东、福建、海南、云南种植。

2. 莞华白

白肉类火龙果（图 2-32），由东莞市林业科学研究所、华南农业大学园艺学院从广州市从化区大丘园农庄引进的红水晶火龙果实生繁殖群体中单株优选而成。果形较整齐，果皮浅红色，厚 0.2 厘

图 2-32　莞华白

米；果肉白色，肉质爽脆、味清甜，可溶性固形物含量 10.8%，单果重 300 克，四年生果树亩产 2 405 千克，适宜在广东中南部地区种植。

3. 仙龙水晶

白肉类火龙果（图 2-33）。由广州仙居果庄农业有限公司、广东省农业科学院果树研究所从白水晶与莲花红 1 号选育，果皮粉红色，厚 0.3 厘米；果肉白色，肉质脆爽，味清甜，可溶性固形物含量 11.2%，单果重 325 克，第五年亩产 2 933 千克。适宜广东中南部地区种植。

图 2-33　仙龙水晶

4. 晶金龙

白肉类火龙果（图 2-34），又称黔白 1 号。由贵州省果树科学研究所用罗甸火龙果园中的晶红龙单株芽变选育而成。果肉白色，近果皮处有红色丝状物，风味清香、味甜。单果重 320 克，第三年亩产 1 685 千克。每年结果 8 批次左右，同批次花

图 2-34　晶金龙

的果实成熟期比其他品种（系）火龙果晚5天。适宜贵州最低温度
0 ℃以上地区种植。

5. 莞华红粉

白肉类火龙果（图2-
35）。由东莞市林业科学研
究所、华南农业大学园艺学
院从红水晶火龙果实生繁殖
群体中单株优选而成。果实
近圆形，果皮浅红色，果肉
白色，近果皮处粉红色，可
溶性固形物含量11.1%。单

图2-35　莞华红粉

果重239克，第三年亩产1 169千克。谢花25～45天果实成熟。
适宜广东中部及南部地区种植。

6. 粤红3号

白中带粉肉类火龙果
（图2-36）。由广东省农业
科学院果树研究所、广州仙
居果庄农业有限公司从白水
晶与莲花红1号选育而成。
花后25～40天果实成熟。
果实圆球形，整齐均匀，果
皮粉红色，厚0.2厘米；果

图2-36　粤红3号

肉白中带粉，肉质细软、味清甜，可溶性固形物含量9.54%。单
果重285克，第五年亩产2 530千克。春梢一般在2月底至5月初
萌发，开花期一般为5月下旬至10月下旬，4批次大花量分别集
中在6月初、7月初、9月初和10月初。果实成熟期在6月底至
12月上旬，较集中的4批次分别是6月底、7月底、9月底和10
月底。适宜广东中南部地区种植。

7. 晶红龙

白肉类火龙果（图2-37）。由贵州省果树科学研究所从普通

白玉龙火龙果中发现的芽变单株选育而成。果实长椭圆形，果皮紫红色，果实鳞片黄绿色、平直。可溶性固形物含量 12.0%，单果重 400 克。四季均能生长，每年结果 7～9 批次，从现蕾到开花 16～18 天，从开花到果实成熟 28～34 天，亩产 1 452 千克。适宜贵州南盘江、北盘江、红水河谷地区种植。

图 2 - 37　晶红龙

8. 双色 1 号

外红内白类火龙果（图 2 - 38）。自花结实，由华南农业大学园艺学院、东莞市林业科学研究所从红水晶火龙果实生繁殖群体中单株优选而成。果实近球形，果皮暗红色，果肉外红内白，肉质软滑、爽脆，味清甜，香味独特，可溶性固形物含量 9.8%，单果重 300～450 克，第三年亩产 1 850 千克。不易裂果，耐贮运。适宜在广东、海南种植。

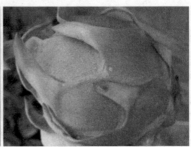

图 2 - 38　双色 1 号

第三节 黄 龙 果

一、黄龙果总述

黄龙果黄皮白肉，也叫麒麟果、燕窝果、黄龙果，鲜食品质佳。花期长，花大而香。从现蕾至开花需 45～60 天，从开花至果实成熟需 90～100 天，在秋冬季则需 110～150 天才可成熟。果实不裂果，自绿转黄需 1 个月（冬季），转黄后可任其挂在树上 1 个多月，是火龙果中品质最佳、口感甜度最好的一个类型。果实具芳香味，果肉中种子大而柔软，可溶性固形物含量 18% 以上，果实略小。主要产果期（秋冬果）在春节前后，夏果则在中秋节上市。

黄龙果分为有刺黄龙和无刺黄龙两个大类，有刺黄龙又分为哥伦比亚麒麟果和厄瓜多尔燕窝果，哥伦比亚麒麟果成熟后果形偏长，厄瓜多尔燕窝果成熟后果形偏椭圆。

二、黄龙果代表品种

1. 厄瓜多尔燕窝果

黄皮白肉类火龙果，又名黄金龙（图 2 - 39）。自花授粉。外形金黄色，因其果肉一丝一丝，状如燕窝而得名，特别爽滑细腻，来自厄瓜多尔，燕窝果品质较佳、口感甜度超好，具芳香味，可溶性固形物含量 18% 以上，有的可以达到 25%。冬果在春节前后、夏果则在中秋节

图 2 - 39 厄瓜多尔燕窝果

上市，燕窝果从花芽冒出至开花需 45～60 天，开花至果实成熟需 90～100 天，秋冬季则需 110～150 天才可成熟。从开花到果实成熟，

需要 120～200 天。单果重 350～450 克，亩产达不到 1 000 千克。种植条件要求严苛，投产时间长达 15 个月。

2. 哥伦比亚麒麟果

黄皮白肉类火龙果，在我国被称为"老黄龙果"（图 2 - 40），需人工授粉。未熟果为绿色，全熟后细刺会脱落，为珍稀品种。花期为每年 6—10 月，开花结果期 160～210 天。单果重 350～450 克。黄色外表形似麒麟，果肉呈透明状，

图 2 - 40　哥伦比亚麒麟果

汁丰润喉，果肉结构呈细丝状，滑如燕窝，可溶性固形物含量在 18％以上，又有些许香味，富含膳食纤维和维生素 C，更含有一般植物少有的植物性蛋白及花青素，含有的铁元素比一般水果高。

3. 无刺黄龙

黄皮白肉类火龙果（图 2 - 41）。又叫黄龙果、黄蜜龙、大黄龙，在我国被称为"新黄龙果"。自花授粉，在大雨天开花成果率也达 90％。果期为每年的 6—12 月。引种于以色列，色泽金黄、耐贮运、个大、无裂果、口感好，

图 2 - 41　无刺黄龙

肉质脆嫩，果肉晶莹通透，水分足，果香味浓厚，甜中带一丝果酸，甜而不腻，助消化。单果重 250 克以上，最大可至 500 克，亩产达 4 000 千克。是目前国内比较耐旱耐寒、抗病率比较高的品种，在高海拔、早晚温差较大的地区种植比较好，适合在广西、广东、福建、海南、云南种植。

第三章
火龙果普通栽培技术

第一节　火龙果繁殖技术

一、扦插育苗

1. 插床育苗

用木箱或直接用地作扦插床，填入干净的河沙（厚度15～20厘米），或用70%的砖末加30%的木炭粉作基质。用砖末加木炭粉作基质效果较好，因为砖末除了有沙的排水透气性能外，本身还能吸附水分，这是沙粒所不具备的性能，所以其具有一定的保水性；木炭粉则具有使繁殖材料预防病菌感染的作用，又具透水透气和保水性能；因而用这两种材料配比出来的基质用于火龙果扦插育苗效果较好。插床准备好以后，剪取30厘米长的火龙果枝条放置阴凉处自然阴干，7～10天伤口愈合后为最适宜的扦插时间，此时扦插可减少伤口感染腐烂，长根发芽率达95%以上。扦插后不能马上浇水，土壤宜干，植后10天才开始第一次浇水。

2. 盆栽育苗

可选用直径17厘米的盆，盆底可铺浮石或砖石，3～7厘米厚，上面铺一层打洞的报纸，再上一层铺蛇纹木屑或树皮木屑，或用3∶1稻壳泥土混合物，约3厘米厚，其上放火龙果枝条，呈斜面种植，上面再放木屑或松树皮6.6～10厘米厚，再放砖头，10天后第一次浇水，以后每3～4天浇水1次，浇水以浇至盆底有水渗出为宜。

二、嫁接育苗

1. 平接法

用尖刀在霸王花（野生三角柱）三棱柱的茎的适当高度横切一刀，然后对三个棱峰以与刀面呈
30°～40°的方向切削，用消过毒的仙人刺刺入砧木中间维管束，将切平的接穗连接在刺的另一端，用刺将接穗和砧木连接起来，砧木和接穗尽量贴紧不留空隙，避免细菌感染不利愈合，然后在两旁各加一刺固定，再用细线绕基部捆紧（图3-1）。

图 3-1　嫁接育苗

2. 楔接法

用消过毒的刀在砧木顶部切开一条裂缝，但不应太深，然后用消过毒的刀片将接穗下部切成鸭嘴形，立即把它插入砧木的裂缝中。用塑料条固定，然后套塑料袋以保持空气湿度，有利于接穗存活。经过20天的观察，嫁接苗若能保持绿色、鲜活，即可成活。1个月后可出圃。

三、播种育苗

取一个熟透的火龙果，取出些果肉，然后浸泡在水里，碾碎果肉，再用过滤布筛洗几遍，去掉多余的果肉，用过滤网继续过滤几遍，直到果肉跟种子完全分离。然后在土中撒下晒干后的种子，用喷壶少量喷水，过3～5天就会发现种子开始发芽，1个月后即可定植。

第二节　火龙果建园

一、选择园地

选择海拔1 400米以下、背风、阳光充足、有便于灌溉和排水

的肥沃壤土、富含有机物的地块或改良的坡地作园地（图3-2）。根据土壤肥力，施用完全腐熟的有机肥2 000～3 000千克/亩，使用拖拉机或微耕机均匀旋转耕作。

图3-2　火龙果园地

二、框架准备

火龙果是茎蔓植物，需要建立架子以便爬藤。火龙果的生产周期通常超过15年，并且框架材料需要强大的耐腐蚀性。通常将10厘米×10厘米水泥桩以200厘米×230厘米的株行距用作框架材料。

1. 单柱式栽培

在距水泥桩顶部10厘米处，以"十"字形钻出两个直径为12～14毫米的孔，该孔用于钢筋穿过后作支撑环。每个水泥桩都需要配备2条直径为12毫米或14毫米、长度为65厘米的钢筋和1个旧轮胎。如果以每亩2米×3米的密度建造桩，则需要准备111个桩。以每亩2.5米×2.5米的密度建造桩，则需要106个桩。

2. 排架式栽培

每亩以株行距（2.2～2.8）米×3.5米的密度建造桩，需要准备70～80根水泥桩。在每根水泥桩距地面高约1米和1.8米处，分别用"十"字形交叉打孔，以备穿钢丝固定。每亩应准备直径4毫米左右的钢丝绳约300米长。

三、移植树苗

1. 单柱种植方法

对于单柱种植，种植密度通常为2米×3米或2.5米×2.5米，并且水泥桩的立栽距离地面以上1.6米，其余应固定在地下。水泥桩立栽后，用土壤固定，然后在水泥桩周围50厘米内施用1.52千克含氮、磷、钾总养分5％的有机细菌肥料，进行混拌，用锄头或旋耕机将它加入 5～10 厘米厚的土壤层中。围绕每个水泥桩在 4 个不同方向挖掘长 8 厘米、深 6 厘米的种植孔，每个孔种植 1 棵幼苗。把育好的火龙果幼苗，每株留一根长的茎芽，带根定植，用土壤覆盖根部，用更耐用的尼龙绳或布条将其绑在水泥桩上，然后倒入生根水。这种栽培方法每亩可以种植 420～450 株，结果枝每亩有 5 000～6 000 条。

2. 排式种植方法

排式种植通常基于株行距 2.8 米×3.5 米或株行距 2.2 米×3.5 米，桩的深度应为地面以上 1.6 米，其余应在地下。每排桩之间的距离为 3.5 米，在 1.8 米处穿孔的桩表面应面向同一方向，并在桩竖起后，钢缆在桩之间通过 1.8 米处的孔将同一排水泥桩连接成一个主体，将火龙果苗以 0.3～0.4 米的植株间距种植在种植行上的两根桩之间，并使用竹条支撑幼苗。当植物在钢缆上生长时，将分枝压在钢缆上，并平行于各行进行分枝，每个植物保留 10～12 个分枝。在 1 米开口处用长 60 厘米、直径为 12 毫米或 14 毫米的钢筋穿过，并在两端使用钢丝以相同方向将同一排水泥桩之间的钢筋拉直。用这种方法每亩可以种植 600～800 株，高产地区的果枝可以达到 6 500～8 500 千克/亩。

第三节　火龙果幼树管理

种植并修剪火龙果幼苗后，应及时补种缺失的幼苗，并在土壤干燥时及时浇水。火龙果成活后，枝条上有许多新芽发芽。此时，每棵植物只剩下一个结实的芽，应及时将多余的芽切掉。

一、种植

火龙果可以单桩种植，也可以行栽。

图 3-3　单桩种植

单桩种植用绳子将新长的茎绑到水泥桩上（图3-3）。当茎长至支撑环位置时，应及时摘心以促进更多分枝的发芽。每株植物留12～15个分枝，并将它们均匀分布在支撑环上。

图3-4 行栽

也可实施行栽（图3-4）。使用绳索将新近生长的茎与水泥柱绑在一起。当植物在钢缆上生长时，及时摘心，可以促进长出新枝条。保持10～12个分枝，使分枝均匀地分布到支撑架的两侧，并除去剩余的芽。

二、肥水管理

火龙果在温暖湿润、光线充足的环境下生长迅速。幼苗生长期应保持全园土壤潮湿。春夏季节应多浇水，使其根系保持旺盛生长状态。果实膨大期要保持土壤湿润，以利果实生长。灌溉时切忌长时间浸灌，也不要从头到尾经常淋水。浸灌会使根系处于长期缺氧状态而死亡，淋水会使湿度不均而诱发红斑（生理病变）。在阴雨连绵天气应及时排水，以免感染病菌造成茎肉腐烂。冬季的园地要控水，以增强枝条的抗寒力。

火龙果同其他仙人掌科植物一样，生长量比常规果树要小。所以施肥要以充足、少量、多次为原则。一至二年生幼树以施氮肥为主，做到薄施勤施，促进树体生长；三年生以上成龄树以施磷、钾肥为主，控制氮肥的施用量。施肥应在春季新梢萌发期和果实膨大期进行，肥料一般以枯饼渣、鸡粪、猪粪按1：2：7配方，每年每株施有机肥25千克。或在当年7月、10月和翌年3月，每株各施牛粪堆肥1.2千克和复合肥200克。

火龙果的根系主要分布在表土层，所以施肥应采用撒施法，忌开沟深施，以免伤根。此外，每批幼果形成后，根外喷施0.3%硫酸镁、0.2％硼砂和0.3％磷酸二氢钾各1次，以提高果实品质。

　　火龙果采收期长，有机肥料要重施，氮、磷、钾复合肥要均衡长期施用。完全使用猪、鸡粪等含氮量过高的肥料，使枝条较肥厚，深绿色且很脆，遇大风时易折断，所结果实较大且重，品质不佳，甜度低，甚至还有酸味或咸味。因此，开花结果期要增施钾肥、镁肥和骨粉，以促进果实糖分积累，提高品质。

　　火龙果的气生根很多，可以转化为吸收根。扩穴改土可逐渐扩宽根系分布；也可绑扎牵引诱导气生根扎根入土。

　　一至二年生的幼树主要以施用生物有机肥或熟化牛粪和氮肥为基础，以勤施薄施来促进树体生长。每亩施用5～8千克的高氮、中磷和低钾混合肥，施用后及时浇水，或每亩施用高氮、中磷和低钾水溶性肥3千克。根据树苗的生长情况，结合浇水，每1～2个月施用一次肥料。

三、栽培要点

　　火龙果的种植方式有爬墙种植和搭棚种植两类。但以立柱栽培最为普遍，即在水泥柱的周围种3～4株火龙果，让植株沿支柱向上生长，每亩可立110条支柱（支柱间距2.5米×2.5米）。按每条支柱种4株果苗计算，每亩可种400多株果苗，大大提高了土地种用率。

　　①加强水肥管理。火龙果一年四季均可种植。因其根系对土壤透气性要求较高，故种植时不可深植。定植前要施足基肥，每穴施腐熟有机肥10千克加复合肥1千克。初期应保持土壤湿润，否则不利于生长。定植后要把土壤踩实，淋足定根水。生产上应"勤施薄施"，开花结果期要增补钾肥、镁肥。冬季遇低温时，要进行灌溉，保持土壤湿润，促使其生长迅速。施肥以有机肥为主，配施少量化肥，在每年3月、7月、10月共施肥3次。每次每株施牛栏粪肥2千克加复合肥0.2千克，促进果实糖分积累，提高品质。

　　②及时摘心、修枝。当枝条长至1.3～1.44米时应摘心，促

进分枝，并让枝条自然下垂，积累养分，提早开花结果。每年收完果后将结过果的枝条剪除，促发新枝，以保证翌年产量。

③ 防治病虫害。幼苗生长期易受蜗牛和蚂蚁为害，可用杀虫剂防治。在高温高湿季节植株易感染病害，导致枝条部分坏死及产生霉斑，防治病害可使用杀菌剂。

第四节　火龙果整形修剪

火龙果属于多年生热带果树，具有生长迅速、萌芽分枝能力强、生殖生长期长等特点，在整个生长过程中，营养生长和生殖生长矛盾突出。

整形修剪作为一项十分重要的火龙果栽培管理技术，涉及整个生长过程。火龙果的整形修剪并非一蹴而就的阶段性工作，而是一项每年都必须根据植株生长结果情况、合理调整枝条分布、更替营养枝与结果枝的日常管理措施。合理适时的整形修剪是实现优质丰产的基础。

一、整形修剪概述

火龙果的整形和修剪是密切联系的，二者互为依靠，整形通过修剪技术来完成，而修剪又是在整形的基础上进行的。

整形就是根据火龙果生长发育特性，以一定的技术措施构建枝条生长的立体空间结构和形态。修剪就是对火龙果的茎、枝、芽、花、果进行部分疏除和剪截的操作。

二、整形修剪的作用

1. 增强通风透光性

火龙果属喜阳性较强的植物，强光照有利于植物开花结果，外层枝条易坐果。但是火龙果长势旺盛，萌枝能力强，极易构建强大的层状树冠，且会快速外移，造成枝条相互遮掩严重，内膛枝条光照不足，不利于营养生长和生殖生长。

通过整形修剪可以合理留存枝条，控制冠形，改善光照条件，增大枝条光合作用叶面积系数，促进粗壮结果母枝形成，为丰产结果打下良好基础。

2. 实现营养枝和结果枝互换

火龙果枝条上的刺座是混合芽，可以萌发出枝芽和花芽，生产上为了提高单果重和品质，保证果实发育的足够营养供给，人为将枝条分为结果枝和营养枝两类。

选择粗壮、饱满、下垂度好、长度适宜的枝条作为结果枝，占枝条总数的 2/3；其余作为营养枝，占枝条总数的 1/3；当年新萌发的预留枝条可以培养为营养枝，翌年也可作为结果枝。

根据生产需要和枝条生长状态，当结果枝上所有萌发的花芽被疏除后就转化为营养枝，营养枝预留花芽后即可转化为结果枝，二者相互替换，提高果树结果性能，有利于稳定果实产量和提高果实品质。

3. 协调好营养生长和生殖生长的矛盾

火龙果的产量多少受坐果期长短、坐果数量多少、果实大小影响。由于火龙果单果发育期短，而全树生殖生长期长，一棵植株甚至一条坐果枝上，大果小果、红果绿果、大花小花、花蕾花芽往往共存，营养竞争矛盾突出。

因此，通过修剪、疏花和疏果等操作，平衡营养生长和生殖生长的矛盾，合理调节果期产量，并形成足够的营养面积，保持中庸健壮树势是获得火龙果树高产优质的关键。

一般盛产期每株果树要保持 18 条以上的枝条，结果枝达 12 条以上，以满足火龙果均衡正常的坐果需要，这一时期每亩要冲施嘉美红利 800 倍液 2～3 次，同时喷施嘉美脑白金 1 000 倍液，促进根系生发和养分运输，促进花果发育。

4. 减轻病虫危害

一方面，经过修剪的火龙果，树势生长强健，增强了机体抗御自然灾害的能力，降低了病虫的侵染概率；另一方面，修剪本身就是疏去病枝、弱枝及残枝，是除病灭虫的基本措施之一。

三、整形修剪的方法

1. 幼树的整形修剪

幼树的整形修剪是为了确保尽快上架，形成有效树冠体系（图3-5）。主要措施是：保留一个强壮向上生长的枝条，以利集中营养、快速上架，当主茎生长达到预定高度后，进行打顶促进分枝，形成树冠立体空间结构。

图3-5　火龙果幼树整形修剪

火龙果定植后15～20天可发芽，平均每天长高2厘米以上，在生长过程中，刺座会生出许多芽苞，前期只留一个主干沿立柱攀缘向上生长，其余侧枝全部剪除，待主茎长至1.5～1.8米的所需高度时，并超出支撑圆盘或横杆30厘米时摘心，促其顶部发生侧枝，一般每枝留芽3条左右，并引导枝条通过圆盘或横杆自然下垂生长，当新芽长至1.5米左右高处时再断顶，促发二级分枝。

上部的分枝可采用拉、绑等办法，逐步引导其下垂，促使早日形成树冠，立体分布于空间。用2～3天时间逐步增加分枝数，最后每株保留枝条15～20条，每个立柱的冠层枝数为50～60条，当枝条数量达到合理设计要求之后，随着侧枝生长，对于侧枝上过密的枝杈要及时剪掉，以免消耗过多养分。

2. 营养生长期的修剪

火龙果营养生长有两个高峰期，主要表现为刺座萌发大量侧芽和茎节增粗。一是在春夏（4—5月）开花结果之前萌发的春枝；二是秋冬（10—11月）开花结果停止后萌发的秋枝。修剪的目的是保持预留枝条总数的动态平衡，适时更新结果枝和营养枝，促进结果枝生长。

　　春枝萌发后，随着光温条件的适宜，树体便进入结果期，因此修剪春枝可以减少养分的消耗，一般情况下如果老枝条预留数量大，结果枝与营养枝配置合理，所有萌芽都应及早疏去，促进枝条尽早进入开花结果期。

　　如果还要配置预备结果枝，可以对老枝条圆盘基部预留侧枝进行培养，其余全部疏去，当新枝条长到 1.5 米左右时及时摘心，每条老枝条最好只保留 1 个侧枝，侧枝总数量以不超过老枝条的 1/3 为宜，这些枝条可以培养成为夏季开花结果的营养枝，翌年春季可作为结果枝。对于病老枝条的更新可配合春枝修剪进行。

　　秋冬季节有大量的新枝长出，一方面要将多余侧芽疏掉，适当在基部留芽培养侧枝，总数以不超过老枝条的 1/3 为宜，以免徒耗营养，新发侧枝长至 1.5 米左右时及时摘顶，促其进行营养积累，这样可以作为翌年春季的营养枝，夏季可替换为结果枝。另一方面，已经坐果较多的当年枝，翌年再次大量、集中开花的可能性较小，在秋冬季结果结束后，应将曾经结过果的全部老枝条剪除，在其基部重新培育大而强壮的秋枝，并随着侧枝的生长和下垂，将其均匀地分布在支撑架的圆盘或横杆上，构建新的结果枝组，以保证翌年的产量。

3. 开花结果期的修剪

　　在生产上如果采用立柱栽培，一般每根水泥立柱可预留 50～60 条下垂枝构成结果枝组，并安排 2/3 的枝条作为结果枝，1/3 的枝条作为营养枝。

　　每年的 5—11 月，果树进入生殖生长期，不间断地分批次开花结果，同时也会从刺座萌发出新枝条，消耗养分，营养生长和生殖生长矛盾尤为突出（图 3-6），为此必须把结果枝和营养枝上新萌发的侧芽

图 3-6　火龙果新萌发的侧芽

全部疏去，减少养分的消耗并促进日光照射，从而满足果实发育的营养需求；同时，还要疏去营养枝上所有的花蕾，缩小枝条生长角度，促进其营养生长，培养为强壮的预备结果枝。

4. 疏花疏果

火龙果花期长，开花能力强，5—11月均会开花，花果盛期，同一枝条上的花苞高达30个，需在出现花苞8天内疏去多余花苞，每枝平均每个花季只保留花蕾3～5朵。授粉受精正常后，可用环刻法剪除已凋谢的花朵，保留柱头及子房以下的萼片即可。

当幼果横径达2厘米左右时开始疏果，原则上每条结果枝留1～3个发育饱满、颜色鲜绿、无损伤和畸形、又有一定生长空间的幼果即可（图3-7），以集中养分，促进果实生长，多余的果、畸形果或病果应及时疏除，这一时期每亩果园要冲施嘉美海力宝4～5千克，同时，喷施嘉美脑白金1 000倍液，为果实迅速膨大和下批花芽发育提供充足的营养。

图3-7　火龙果的幼果

四、注意事项

1. 修剪位置的确立

枝条的长度、数量和下垂角度设置得科学合理是取得高产的基础，整形修剪的目的就是构建有利枝条生长的良好空间结构体系，包括主干的确立、营养枝和结果枝长度的设置，以及枝条的下垂分布等。

结果枝长度一般大于1.5米，中上部的枝条、枝条顶端和下垂枝最容易结果，而中下部的枝很少开花，上部枝条生长势通常大于中下部枝条，这可能是受到顶端优势影响。因此，无论是摘心还是培养新枝条，都要掌握好合适的位置和长度，要引导枝条下垂，不可盲目操作。

2. 把握好枝条生长的有序性

营养生长是生殖生长的基础，主要表现为茎节增粗、分枝数量增加及延长生长，营养枝数量和质量不仅关系果实产量和品质，而且关系结果枝的替换和产量稳定。因此不能一味地追求结果数量而忽视营养枝的配备，使所有枝条都开花结果，这种枝条培养的无序性对定量栽培管理技术的应用是不利的。

3. 注意修剪质量

无论是修剪枝条还是疏芽都应在晴天有太阳照射下进行，以便树体伤口愈合，避免病菌侵入。修剪刀具要锋利，修剪动作要利索，还要避免损伤枝条。修剪时所有用具应提前用酒精或者高锰酸钾消毒。修剪、疏芽、疏花和疏果要及时进行，切忌多余树体组织或器官过分消耗营养后再操作。

第五节　火龙果人工授粉

有些火龙果品种特别是红肉火龙果，雄蕊与花柱会出现等长或者雄蕊短于花柱的现象，故自花授粉坐果率较低，加上火龙果夜间开花没有昆虫帮助授粉，故需进行人工辅助授粉。

火龙果不同品种遗传差异越大，彼此间授粉率会越高。因此可以适当间种一些不同品种的火龙果，尤其可以混种红肉、白肉品种。

一、授粉时间

要根据火龙果的开花规律来判断授粉时间。火龙果一般都是在下午 6 时开花的，开花时间也不长，午夜 12 时左右花冠颜色开始变淡，基本上开花时间只有一个晚上左右。因此，火龙果人工授粉应在每天傍晚花开到早上花朵逐渐闭合的时候进行，最好能够在凌晨1时进行授粉，此时火龙果花朵的花径是最大的，授粉成功率也最高。

二、授粉方法

授粉方法可使用最简单的毛笔点授法。因为火龙果的花朵是比较大的，而且每朵花上的花粉都是比较多的，因此花粉采集起来也非常简单。不过要注意，花粉的活力是比较差的，在常温环境下，花粉存活的时间一般不会很长。因此在进行人工授粉时，花粉最好随采随用，以保证花粉的活力及授粉效果。花粉采集好之后，直接用毛笔蘸取花粉，然后将其均匀地涂抹在柱头上。

也可以先将容器套在火龙果的花蕾上，轻拍花朵，花粉就会落到容器中。再用毛笔、刷子等有毛状物工具，蘸取采集到的花粉，抹到雌蕊上。一般，授到柱头上的花粉量越多，火龙果果实就会越大（图3-8）。

图3-8　授　粉

授粉1~2天后，要将花瓣剪掉，避免花瓣太厚、保湿效果太好造成腐烂。开花期遇下雨天，要把花苞罩起来，授完粉也罩上。

第六节　火龙果病害防治

火龙果有肉质茎，花朵肥大，果实多糖，生长过程容易滋生各种微生物，尤其是火龙果还生长在高温多湿环境中，因此病害发生更为严重。

一、火龙果茎枯病

该病是火龙果的主要病害，一般在 3 月下旬开始发病，此期病害开始见病，但病株率尚低，病情发展缓慢。从 4 月上旬开始，病害开始加重。5 月中旬至 7 月中旬为发病盛期。高温高湿有利于病害的发生，全年发病高峰时间是 6 月底至 7 月初。11 月下旬以后进入越冬阶段。

二、火龙果软腐病

水分与软腐病发病的关系最为密切。多雨潮湿地区或土壤水分过大，有利病菌的繁殖、传播和蔓延，造成该病的暴发流行。

温度也是影响火龙果软腐病发生的极重要因素。温度过低时，伤口不易愈合，为病菌侵染创造了有利条件。一般从 10 月就开始发病，翌年 1 月下旬至 3 月上旬是发病盛期，4 月气温回升时病情减轻。该病具有发病急、蔓延快、危害大的特点。

三、火龙果茎斑病

主要危害火龙果肉质茎，高温高湿最有利于该病的发生，其中高湿是发病的主要原因；其次，枝条过密、通风透光差、偏施氮肥以及培肥管理差的火龙果园发病较重。

四、防治技术

火龙果主要病害的防治应采取以农业防治为主、化学防治为辅的综合防治措施。

农业防治。建立无病苗圃，选择无病种苗，保护无病区；种植和选育抗病新品种；清除并集中销毁病残体，清除杂草，降低田间湿度，减少田间病源；加强肥水管理，增施磷肥、钾肥，提高植株抗病性。在火龙果的 3 种主要病害中，由于火龙果软腐病属细菌性病害，发病快，不易防治，因此，多采用农业防治措施，如杜绝用带病苗种植、及时清除病残体，以减少病菌的繁殖、传播和蔓延。

化学防治。在发病初期及时修剪和刮除腐烂部位，对伤口喷施杀菌剂。火龙果茎枯病应在发病初期用50％福美双可湿性粉剂500～800倍液或50％甲基硫菌灵500～800倍液，每隔7天喷1次药，共喷3次，对防治火龙果茎枯病有良好效果。

第七节　火龙果虫害防治

为害火龙果的害虫主要是刺吸式口器昆虫、蜘蛛。

一、堆蜡粉蚧

该虫主要为害新梢，附着于茎棱边缘，在光照不足或阳光照不到的茎蔓常有发生，以刺吸式口器插入茎肉吸收营养。

防治方法：用小竹棍绑上脱脂棉或用小棕刷刷去粉蚧，集中灭杀；用尼古丁、肥皂水洗刷；除火龙果植株开花期外，采用喷洒浇水的方法，可防治堆蜡粉蚧的侵害。

二、黑刺粉虱

黑刺粉虱属同翅目、粉虱科。此虫主要为害火龙果茎，从茎中吸食汁液，影响植株生长。在火龙果茎尖、棱的边缘处常有白粉状黏附物，开始是小白点，以后逐渐扩大。

防治方法：农业防治采取增施有机肥、配施氮、磷、钾肥，适时修剪等措施，不断改善火龙果的肥水和通风透光条件，以增强树势，提高抗虫能力。对于越冬害虫的防治可分别于3月上旬、10月下旬用45％石硫合剂结晶150～200倍液进行防治，铲除越冬害虫的幼虫和卵，降低大田虫口基数。盛发期可选用2.5％联苯菊酯2 000倍液、吡虫啉3 000倍液喷药防治，间隔10天连续喷药2次，可达到良好的防治效果。

三、红蜘蛛

红蜘蛛属蛛形纲、蜱螨目、叶螨科。分布广泛，食性杂，可为害110多种植物，在火龙果园中主要为害火龙果植株新梢，留下丝状物。雌成螨深红色，体两侧有黑斑，椭圆形。越冬卵红色，非越冬卵淡黄色。越冬代幼螨红色，非越冬代幼螨黄色。越冬代若螨红色，非越冬代若螨黄色，体两侧有黑斑。

防治方法：一旦发现火龙果植株上有红蜘蛛，要马上用药，可选用克螨特、哒螨灵、乙螨唑、联苯肼酯和阿维菌素等农药。每隔5～7天喷1次药，连用2～3次。最好在黄昏时用药，要喷透、不留死角。采用轮换用药或混合用药的方式可取得很好的防治效果。

第八节　火龙果采收

火龙果授粉成功的花朵，授粉后30～40天，果皮开始变红，且有光泽出现时即可采收（图3-9）。

火龙果宜适期采收，过早或过迟采收均有不良影响。过早采收，果实内营养成分还未能转化完全，影响果实的品质和产量；过迟采收，则果质变软，风味变淡，品质下降，不利运输和贮藏。采收前几周应制定好采收计划，做好一切准备工作，应先熟

图3-9　采　收

先采，分期采收，供贮藏的果实可比当地鲜销果实早采，而当地鲜销果实和加工用果，可在充分成熟时采收。

火龙果应在适宜的天气采收，最好在温度较低的晴天早晨、露水干后进行。雨露天采收，果面水分过多，易使病虫滋生。大风大雨后应隔2～3天采收。晴天烈日下采收，则果温过高，呼吸作用旺盛，降低贮运品质。

采收时用的果剪，必须是圆头，以免刺伤果实。果筐内应衬垫

麻布、纸、草等物，尽量减少果实的机械损伤。

采收时，用果剪从果柄处剪断，轻放于包装筐或箱内即可（图3-10）。采收时要尽量保留果梗，带有果梗的果实在贮藏过程中比不带果梗的果实重量损失少，其成熟过程慢一些，贮藏寿命也就相对长一些。保留果梗时可用果剪齐蒂将果柄平剪掉，

图3-10　贮　藏

这样可避免包装贮运中果实相互划伤。

采收后的火龙果果实应放在阴凉处，不能日晒雨淋，采收后进行果实初选，按果实的大小和饱满程度分级包装，果实经挑选、分级、清洁后，用纸箱或木条箱盛装，逐个放在箱内将果实固定，分层叠放，这样可大大减少果实在贮运中所受的机械损伤，也可提高果实的商品档次。

一般水果如呼吸率高则贮藏期短，呼吸率低则贮藏期长。采收后的火龙果属于呼吸率低的水果，果皮厚且有蜡质保护，极耐贮运。在常温下可保存15天以上，若装箱冷藏，贮藏温度在15℃左右时，保存时间更可长达1个月以上。但在高温季节，火龙果收获后必须置于阴凉处散热，并进行冷藏以利于保鲜。火龙果的这种易贮藏特性，在市场销售中更具竞争性和便利性。

第四章
火龙果现代化栽培技术

为了提高火龙果果实的产量和质量，拓展火龙果生产范围，提高火龙果生产效率和效益，应采用搭棚、补光和水肥一体化灌溉等现代化生产技术。

第一节　火龙果设施栽培技术

火龙果为热带水果，引到北方种植，必须通过必要的设施为其生长创造适宜的环境条件。因此，要选择保温效果好、采光充足的温室种植。北方寒冷地区常用的栽培设施有日光温室、单栋拱棚、连栋大棚等，夏季用遮阳网降温，冬季覆盖加厚防寒帘或保温被，或者搭建内棚，温室内设加温炉，以应对阴天、雨雪等严寒天气。

一、设施条件

采用连栋温室大棚，棚高 2.8 米，棚宽 6.5 米，大棚钢管均为径粗 20 毫米、壁厚 2 毫米的镀锌钢管，夏季单层膜覆盖，棚膜为厚 10 毫米的白色塑料膜，冬季低温季节加盖内膜，实施双膜覆盖保温，内膜规格为厚 4 毫米的白色塑料膜。大棚建设基本以南北向为主（图 4-1）。

图 4-1　火龙果设施栽培

棚内以耐湿、耐高温、耐腐蚀、支撑力强的钢筋水泥柱作为火龙果攀缘支撑物,规格为 8 厘米×8 厘米×180 厘米(长),水泥立柱 30 厘米深埋地下固定,地上保持 150 厘米,并以径粗 12 毫米钢丝牵拉固定,以在火龙果生长时攀缘开花结果时用。

安装自动卷帘机、微喷灌、室内照明等设施。每棵植株旁都有一个滴灌孔,实现肥水一体化供给,确保水肥及时调节和供应。在大棚内为充分利用空间,在操作施肥、起沟过程中,每棚做 3 垄,垄高 20~25 厘米,两垄间留 30 厘米宽的沟,以作田间作业道路和灌溉用。

二、主要技术

1. 水分管理

一般生育期保持土壤含水量 70%~80%,开花期保持空气相对湿度 60%~70%。定植初期,每 2~3 天浇 1 次水,幼苗期和果实膨大期保持土壤湿润,以利果实生长。干旱时每 3~4 天灌 1 次水即可;连阴雨天及时排水,以免植株感染病菌造成茎肉腐烂。

2. 温度管理

火龙果在 17~25 ℃时有利于植株生长,昼夜温差 10 ℃以上有利于花芽形成和正常授粉,25~38 ℃有利于果实发育。为保证火龙果常年生长和多次结果,尽量满足上述温度条件,维持生长适宜温度 25~35 ℃,做到冬季夜温不低于 8 ℃,夏季高温期注意通风。

3. 光照管理

火龙果需要较强的光照,一般要求光照强度在 8 000~12 000 勒克斯。若光照强度不足,特别是生殖生长期,遇到 8 000 勒克斯以下的光照强度,设施内都要采取不同程度的补光措施;若光照过强,则要适度遮阴。为实现火龙果的周年开花结果,一般都采用补光措施。

第二节 火龙果补光技术

火龙果是一种热带水果,在海南等热带地区,5 月至 12 月中旬能自然开花结果，12 月末至翌年 4 月不能自然开花结果。为了延长结果期，可以通过冬季补光的生产方式实现增产。其原理是人为模仿阳光照射，促进火龙果进行光合作用诱导成花（图 4-2）。

图 4-2 补 光

火龙果花芽产生与花芽分化的临界日照时间接近 12 小时，日照短于 12 小时开始进行补光，也就是在当年进入秋分之后至翌年春分之前，用人工光源来满足火龙果对日照时长的需求，可以诱导火龙果产生花芽，提高反季节产量。

一、补光灯选择

选择植物补光灯时，可用 LED 灯，比较节能（图 4-3）。补光灯一般是蓝红光结合的黄光，蓝光波长 430 纳米左右，可以促进果实胡萝卜素、维生素形成，红光波长 630 纳米左右，可以诱导开花结果，

图 4-3 LED 补光灯

红蓝光相结合波长在 510～610 纳米之间，色温 3 000～4 000 开尔文，每平方厘米照射 120 勒克斯，功率 12～18 瓦。

二、补光灯悬挂位置

补光灯要挂对高度、悬挂均匀。悬挂高度为距植株顶部 50 厘米左右，太高了补光亮度达不到，太低了容易被枝条遮挡。目前火龙

果一般采用连排式栽培，每亩栽培1 000～1 500株，悬挂灯间距为1～1.5米，具体可根据补光灯功率来定，应使主要结果枝补光强度达120～150勒克斯。

三、环境温度

温度也是火龙果开花结果的一个关键因素，当温度低于15 ℃，怎么补光都很难开花结果，温度在15 ℃以上可进行补光生产，温度在20 ℃以上补光效果最佳。

四、补光时间

为满足火龙果对日照的需求，同时节约能源，目前田间生产多采用每日照灯4小时的方法。一般补光时间有：18:00—22:00；22:00—02:00；02:00—06:00三种选择。实际做法其实是可以依据每周日长变化，逐渐调节照灯时数，亦即每周照灯时间2～4小时不等，如此可节省照灯时间，降低用电成本。

像云南、贵州等地区，补光时间一般以秋分和春分为界，秋分过后开始挂灯补光，直到翌年春分（图4-4）。像海南那样的热带地区，温度较高，每年10月中旬至3月下旬补光。每天补光2～4小时，日落后就补，这时可借助

图4-4　日落后补光

太阳余温，补2小时左右的光就够了，开始补光时间越晚，则需要补光时长越长。

第三节　火龙果水肥一体化技术

农业是经济发展的支柱行业，未来节水仍是农业发展的主要方

向，我国仍会加大投入以推进节水灌溉技术的发展。水肥一体化系统的推广和应用，将会有效改善我国农业灌溉的落后面貌，加快我国农业灌溉现代化步伐，为绿色、生态、高质量的农业发展贡献一份重要力量（图4-5）。

图4-5　水肥一体化系统

　　一套完整的水肥一体化系统，想要保证管道不堵塞、灌溉系统完整运行，除了核心水肥一体机，过滤器也是必不可少的一部分。水肥一体化系统中常用的过滤器有离心过滤器、砂石过滤器、网式过滤器、叠片式过滤器等；除了过滤器还有电磁阀、灌溉管道、灌溉喷头等一系列配件。

　　采用物联网（IOT）智能水肥一体机作为核心控制中心，实现定时定量、科学灌溉。在满足作物特殊的水肥需求的同时，又能增加作物产量，提升作物品质；同时还能开源节能，应用水肥一体化技术平均节水50%、节肥40%、省工90%，方便园区管理，提高工作效率。

　　滴灌将水流均匀渗入植物根系附近土壤，滴水流量小，水缓慢进入土壤，可最大限度减少水分蒸发损失，水源利用率可达95%以上。

　　相比滴灌设备和喷灌设备，雾化微喷设备具备一些二者皆不具备的特点，可以应用于某些不适合滴灌和喷灌的场景，达到二者都不能达到的效果，是一种应用潜力较大的节水灌溉方式（图4-6）。微喷的特点是喷洒水量小、灌溉范围大、平均用水少，微喷头灌溉系统压力低，不需要大型加压设备，能耗较小，将施肥罐中的水肥溶液经过网式过滤器过滤以后，再通过微喷头喷洒而出，同时对作物进行根部施肥与叶面施肥，提高了肥料利用率。采用折射雾化微喷头灌溉比传统的地面灌溉省水70%以上，节省肥料30%以上，比用大喷头、中喷头灌溉节能50%左右。

图 4-6 微 喷

参考文献

蔡永强，2017. 火龙果栽培关键技术［M］. 北京：中国农业出版社.

樊军，2015. 陕西澄城火龙果设施栽培关键技术［J］. 果树实用技术与信息（12）：2.

刘友接，2019. 火龙果优良品种与高效栽培技术［M］. 北京：中国农业科学技术出版社.

马玉华，蔡永强，2017. 火龙果高效栽培技术［M］. 贵阳：贵州科技出版社.

秦永华，等，2020. 火龙果优质丰产栽培彩色图说［M］. 广州：广东科技出版社.

孙诚志，2005. 火龙果设施栽培技术［J］. 广东农业科学（2）：2.

Chapter I

Biological Characteristics
of Dragon Fruit

Section Ⅰ   Overview of Dragon Fruit

Ⅰ. The Family and Genus of Dragon Fruit

Dragon fruit, also known as Honglongguo (Red Dragon Fruit) or Xianmiguo (Fairy and Sweet Fruit), is a perennial climbing succulent plant of the *Pterodactylus* genus of the Cactaceae family. The dragon fruit is native to Central and South America, and is distributed from Central America to northern South America, and is widely cultivatecl throughout the world.

Ⅱ. The Nutritional Value of Dragon Fruit

With its oval shape, its red or yellow color, the dragon fruit is of unique appearance. It is bright in color, rich in nutrition, and unique in flavor. Its characteristics of low-calorie, high-fiber, high-nutrition, functions of detoxifying and relieving the internal heat have won it the reputation of 'Queen for Cosmetology' and make it very popular in the market.

The flower of dragon fruit is not only rich in amino acids, trace elements and etc. , but also contains rich elements that have factional effects. Therefore, it is of unique functions.

The dragon fruit is rich in nutrition, sweet and juicy, rich in vitamins, glucose and minerals such as potassium, calcium and magnesium which are important to the human body. It contains plant albumin, anthocyanins, rich Vitamins C and water-soluble dietary fiber which are rare in other plants. Albumin is a viscous and colloidal substance, which serves as a detoxifier for heavy metal poisoning and can protect our stomach wall. The dragon fruit contains relatively rich anthocyanins, which is especially high among the red fruit. The anthocyanins have anti-oxidation, anti-free radical, anti-aging effects, and can also enhance the prevention of brain cell degeneration and prevent dementia, while Vitamin C has skin-whitening effects. Lastly, the rich water-soluble dietary fiber helps to lose weight, reduce blood sugar, moisturize the intestines, and adjunctly prevent colorectal cancer.

The dragon fruit seed oil contains as high as 78.10% unsaturated fatty acid, among which the monounsaturated fatty acid accounts for 33.39%, while polyunsaturated fatty acid accounts for 44.71%, linoleic acid up to 44.29%, which is even higher than rapeseed oil and peanut oil and close to sesame oil, and oleic acid as high as 31.75%, which is a monoethylenically unsaturated fatty acid.

The skin and flesh of dragon fruit contain a large amount of red pigment, which is a good source of natural edible red pigment that can be extracted and processed. The red pigment in its skin is water-soluble and alcohol-soluble, stable under acidic conditions. It shows bright red color with $pH = 1.5$, and is soluble in ethanol, acetone, isopropanol, acetic acid, citric acid, tartaric acid and 0.2 mol/L HCl solution, slightly soluble in ether and ethyl acetate, presenting the characteristics of polar or weakly polar molecules.

Ⅲ. The Economic Value of Dragon Fruit

The dragon fruit is an edible and ornamental tropical fruit. It can be planted in pots or in fields all year round and has multiple flowering and fruiting cycles with high yield.

Taiwan is the first region in China to introduce some varieties of dragon fruit, while Guangdong is the first province in China's mainland to plant the dragon fruit.

At present, the dragon fruit is mainly cultivated in the five major production areas in Guangdong, Guangxi, Hainan, Yunnan, and Guizhou, and Nanning, Guang Xi owns a dragon fruit planting area of 150 000 mu*, which makes it the largest dragon fruit production base in China. It is also planted in Fujian, Liaoning, Sichuan, Shanghai, Zhejiang and other places.

Section II   Biological Characteristics and Characteristics of Growth of Dragon Fruit

I . Biological Characteristics of Dragon Fruit

1. Categories

Dragon fruit (*Hylocereus undulatus* Britt) refers to dicotyledonous class, Cactaceae order, Cactaceae family, Columnar Cactus subfamily, and *Pterodactylus* genus (Fig. 1.1).

Fig. 1.1   Dragon Fruit

At present, on the fruit market, the red dragon fruit, purple dragon fruit and golden

---

* Mu is a non-statutory unit of measurement, 1 mu＝1/15 ha, it's the same for the rest of the book. ——Editor's note

dragon fruit are collectively referred to as dragon fruit. Red dragon fruit and purple dragon fruit have a red skin and red or purple flesh, while golden dragon fruit has a yellow skin and white flesh. Although collectively referred to as dragon fruit, they are plants of the same genus but different species. The scientific name of red dragon fruit is *H. polyrhizus*, the scientific name of purple dragon fruit is *H. costaricensis*, and the scientific name of golden dragon fruit is *H. megalanthus*. There are about 20 species in this genus, which grow in Central America and Mexico. The morphological characteristics of the main varieties and their propagation and cultivation methods will be introduced as follows.

Dragon fruit, an epiphytic climbing plant, is able to yield fruit by self-pollination. Warm and humid environment is conducive to its growth. The optimum temperature for growth varies from 25 ℃ to 35 ℃. Intolerant to low temperature, its stem nodes are susceptible to be contaminated by disease and rot when the temperature is below 3 ℃. It likes fertile loam land with much humus, and the potting soil can be prepared with equal amount of leaf humus, coarse sand and decomposed manure. Vileplume is mostly propagated by cutting. During the growing season, the full-grown or older stem nodes are cut, and the cuttings should be no shorter than 15 cm. After cutting, air the cuttings for a few days. After the incision becomes dry, insert it into the sand bed or directly into the soil. It can take root after one month. Transplant the cuttings into a small pot or directly cultivate them in the open field when the root length grows to 3~4 cm. During the growing season, the soil must be fully irrigated and top dressed with decomposed liquid fertilizer once every half month. Limited irrigation and no fertilization in winter. The dragon fruit can be planted in pots as ornamental plant. However, for fruit production, large field cultivation is recommended.

The dragon fruit is a tropical plant that can grow rapidly in warm and humid environment with proper sunlight. It can withstand low temperatures down to 0 ℃ and high temperatures up to 40 ℃. To ensure perennial growth and multiple yields, it is important to try to create a conducive environment with temperatures between 25 ℃ and 35 ℃. When cultivating in the open field in spring and summer, it should be watered more to maintain the vitality of the root system. On rainy days, the land should be drained in time to avoid the infection of bacteria which will cause the stem and the flesh to rot. The dragon fruit is also adaptive to different kinds of soils, but neutral soil and slightly acidic soil that have rich humus, retain water and fertilizer well are the best. Apply more fermented organic fertilizer as base fertilizer, apply more calcium, magnesium, phosphorus and compound fertilizers at the seedling stage, add potassium and magnesium fertilizers during the period of flowering and fruiting, to ensure vigorous growth, sugar accumulation and improve fruit quality. The dragon fruit begins to bloom and bear fruit 12 ～ 14 months after planting. In the second year after planting, each plant produces more than 20 fruit, and in the third year it enters the full bearing period. During the cultivation of dragon fruit, attention should be paid to topping to promote branching. Let the branches droop naturally, hence, it can bear more fruit and would be easy to harvest.

Red dragon fruit has multi-branches up to 2 m long, with 3-edged stems, sometimes 2 or 4-edged. The stems are dark green, with areoles on the edge, and there are 2～6 short spines on each areole. The flowers are large, white, 25 ～ 30 cm in diameter, bloom at night. It is difficult to bear fruit in natural environment.

Pollenating by dragon fruit's pollen can increase the number of fruit, however, the flesh remains red. The market price of red dragon fruit is higher than that of dragon fruit, but the yield is slightly lower.

Purple dragon fruit's stem is 3-edged, with areole on the edges, and there are 2~4 short spines on each areole. The flowers are white, fragrant, about 30 cm in diameter. It is also difficult to bear fruit in natural environment. Pollenating by dragon fruit's pollen can increase the number of fruit, but the flesh remains purple-red. A small amount of purple dragon fruit has been cultivated in China.

Golden dragon fruit is probably the plant with the largest flower in the cactus family. The plant is epiphytic, climbing and crawling. The stem length can reach 3 m, with 3 slightly curled edges. The areoles are white, with 1~3 short spines. The flowers are white, fragrant, 32~38 cm in diameter. The skin is yellow, and the flesh is white, with great taste. A small amount has been imported to the domestic fruit market, but it has not been cultivated in China.

2. Flower

The flowers of dragon fruit are hermaphroditic, generally one flower for one fruit. The buds are funnel-like, the calyx is yellow-green, and the flesh is thick and scale-like. The buds mostly flower at night, the petals are pure white, while a few species have red petals. The stamens are slender and numerous, the pollen is milky yellow, and the stigma of the pistil is green. The buds are more than 30 mm long a before blossom with pure white petals. The buds are yellow-green, smooth and huge. The flowers are 30 cm long. Hence, it is known as "Dawanghua (Big Flower King)" (Fig. 1. 2).

Fig. 1. 2   The Flowers of Dragon Fruit

3. Fruit

The shape of the dragon fruit's fruit is unique and has great ornamental value: the fruit is spherical and of elegant shape. The color gradually changes from green to red (except for the yellow-skinned species) during growth, and the color is bright red or rose when mature. The weight of a single fruit is 350～500 g. Most of the fruit are red and edible. The time from flower bud differentiation to flowering is relatively short, generally 40 ～ 50 days. Self-pollination is with low fruit setting rate, and artificial pollination is recommended. Dip the pollen with your fingers and put it on the stigma. 25～30 days after pollination, the color of the skin turns red, and the fruit can be harvested 4～6 days after the skin turns red. The plant enjoys long longevity and can live up to more than 100 years. Within a certain period of time, the yield increases with the aging of the tree.

4. Root

The dragon fruit is a succulent plant with a shallow root

system. It's not necessary to dig a pit when planting. Insert the root 3 cm deep and moisturize the soil, mix the basal fertilizer, and insert a climbing column for the growth of the plant. The survival rate can raise to 100%.

5. Stem

The nodular stem of this genus usually has only 3 edges, the main stem can generally produce 8 main branches, with a diameter of up to 13~18 cm. The stem nodes have climbing roots, and each stem node has 1~3 short spines in the recessed part. It crawls or climbs, growing to more than 10 m long, with many aerial roots on the stem. The areoles are sparse, the spines are cone-shaped and very short. The stem has two or three edges, the leaves are dark green, and the edges are wavy with small spines. The branches and leaves are integrated with the shape of a sword.

6. Growing environment

The dragon fruit is a tropical fruit that likes light and resistant to shady environment, heat and drought. It likes fertile land but can be grown in barren land. It grows rapidly in warm and humid environment with proper sunlight. When cultivating in the open field in spring and summer, it should be watered more to maintain the vitality of the root system. On rainy days, it should be drained in time to avoid the infection of bacteria which will cause the stem to rot. Its stems can also develop when grows onto rocks, and the plants are extremely wind-resistant. As long as the climbing pole is firm, it can resist typhoons. It can withstand low temperatures down to 0 ℃ and high temperatures up to 40 ℃. The optimum temperature for growth is 25~35 ℃. Open field cultivation should be avoided in areas where temperature might drop below 8 ℃ in winter. The dragon fruit can be planted in different types of soil. However, neutral soil and slightly acidic soil that contain more humus, with good water and fertilizer retention and drainage suit it best.

Ⅱ. Characteristics of Growth of Dragon Fruit

1. Strong Adaptability

The appearance of dragon fruit is very similar to Vileplume (Wild Trigonocarpus). The dragon fruit can be planted anywhere that the wild Trigonocarpus can grow. It can be planted on flat land, slopes, paddy field and dry land. Cactaceae plants are tropical plants, so they should only be planted in south areas of the Tropic of Cancer in China. A long frost period will affect its growth.

2. Few Problems of Pest and Disease

The dragon fruit is a kind of fruit tree with few disease and pests. In the seedling stage, large ants may be attracted to bite the seedlings due to the application of farmyard manure. However, they can be removed easily with ant termites. When it grows up, especially from flowering to fruit ripening, no pesticide is required. In the production area, almost all other fruit need to be bagged, while the dragon fruit is the only exception.

3. High Yield

The cultivation technique of dragon fruit is relatively simple, and the yield of orchard over 3 years can reach more than 825 kg per mu. Judging from the trial planting of our farm, such yield can be achieved easily. Plant 110 plants per mu with individual plant yield of 7. 5 kg per mu. Generally, the weight of a single fruit is more than 0. 25 kg, and each plant has more than 40 fruiting branches. Each fruiting branch has several flower buds, and the redundant flower buds need to be removed manually.

4. Long Fruit Bearing Period

The dragon fruit has high-yield. Simultaneously, the fruit-bearing period can last as long as 180 days, and it matures in different cycles to ensure that the fruit could be on the market in different times. Hence, the farmers will have a stable income. According to the trial planting, it takes about 40 days from flowering to harvest, and the fruit will not fall off by itself from the branches even it is fully ripe.

5. Easy to be Stored and Transported

After the dragon fruit is harvested, it can be stored at room temperature for $14 \sim 21$ days, while can remain fresh for several months under refrigerated conditions. As the skin is thick and protected by wax, it is not easy to be damaged during transportation and easy to be stored and transported. Within the shelf life, the red color of the skin becomes increasingly condensed.

Chapter II
Varieties of Dragon Fruit

There are three subspecies of dragon fruit: red-fleshed subspecies, white-fleshed subspecies, and yellow-fleshed subspecies. At present, there are mainly three varieties of dragon fruit in China: red-skinned with white flesh, red-skinned with red flesh, and yellow-skinned with white flesh. Additionally, there are green-skinned with red flesh, two-colored dragon fruit, etc.

Section I   Red Dragon Fruit

I. Overview of Red Dragon Fruit

The red dragon fruit has red or purple-red flesh (Fig. 2. 1). It tastes great when consumed fresh. The fruiting period is from May to November in Taiwan, Guangdong and Guangxi provinces, while it is longer in Hainan province. The self-pollination rate and setting rate of red-fleshed dragon fruit are low. However, the fruit setting rate is higher and the fruit is larger when pollinated by white-fleshed dragon fruit.

The fruit can be processed into juice, fruit powder, red pigment, ice cream powder, jelly and jam.

The red-fleshed dragon fruit is oval, round, or cylindrical in shape, with bright red and shiny skin, red or purple-red flesh. It is delicate and juicy, with $16\% \sim 21\%$ soluble solids and high sweetness. It contains a large amount of anthocyanin, which is an antioxidant whose antioxidant capacity is more than 10 times stronger than that of carotene. The weight of a single fruit is 350~1 000 g, and the yield per mu is 1 000~ 3 000 kg.

Fig. 2. 1 Red Dragon Fruit

II. Representative Varieties of Red Dragon Fruit

1. Fenhonglong (Pink Dragon)

Pink-fleshed dragon fruit (Fig. 2. 2). It is selected and bred from bud mutation of individual plant of Xinghonglong (New Red Dragon) by Guizhou Provincial Institute of Fruit Tree Science. Its fruit is oval with red skin and pink flesh, and the skin is 0. 29 cm thick. It

Fig. 2. 2 Fenhonglong (Pink Dragon)

contains betalain and has black seeds. The average weight of a single fruit is 340 g, and the yield per mu is 1 834. 7 kg. The best regions for its cultivation are areas in Guizhou Province where the annual average temperature is above 18. 5 ℃ and the temperature in January is above −1 ℃.

2. Jinzuan (Gold Drill)

Red-skinned with red flesh dragon fruit, which is also known as Nvwangtou (Queen's Head) (Fig. 2. 3). It is a self-pollinated

Taiwan variety. The weight of a single fruit is over 800 g, and the maximum is up to 2 500 g, with the soluble solid content of 17%~20%. Maximum yield can achieve over 4 000 kg per mu. Flowering begins in April, and the interval between each flowering cycle is about 15 days. The flowering period ends at the end of October.

Fig. 2. 3   Jinzuan (Gold Drill)

Totally, there are about 18 flowering cycles. The fruit-picking period is from late June to early December, and 8 cycles of fruit can be picked each year. It takes 30~40 days from flowering to harvest. The best regions for its cultivation are Guangxi, Guangdong, Fujian, Hainan, and Yunnan provinces.

3. Xianglong (Propitious Dragon)

Red-skinned with red flesh dragon fruit (Fig. 2. 4). It is a self-pollinated Taiwan variety. The average weight of a single fruit is 418. 6 g, and the maximum weight of fruit can be more than 1 000 g. Its soluble solid content is 16. 5%. In the forth year, the transplanted

Fig. 2. 4   Xianglong (Propitious Dragon)

seedlings enter the full fruit period, and the yield per plant is more than 8 kg. Buds appear in mid-April, and the flowering period lasts from late April to early and mid-November. The fruit-picking period is from late June to early December, and 8 cycles of fruit can be picked each year. It takes 35~40 days from flowering to harvest. The best regions for its cultivation are Guangxi, Guangdong, Fujian, Hainan, and Yunnan provinces.

4. Xixianghong (Happy, Fragrant, Red)

Red-fleshed dragon fruit (Fig. 2.5). It is a self-pollinated Taiwan variety, and the weight of a single fruit is 500 g. The yield per plant is 6.2 kg. Its fruit is round, with bright red skin, blood-red flesh. It is rich in anthocyanins. It tastes tender and extremely sweet with high-sugar content. It is juicy and sweet with soluble solid content of 18%~22%. It can be planted in orchards. Its thick skin makes it easy to store and transport. The first flower buds appear in mid-April and the last batch of

Fig. 2.5   Xixianghong (Happy, Fragrant, Red)

flowers wither in early December. The seedlings are generally planted for 10~12 months, and the branches can bloom and bear fruit when the branches grow higher than 1 m. The fruiting period is from May to November every year. The best regions for its cultivation are Guangxi, Guangdong, Fujian, Hainan, and Yunnan provinces.

5. Taiwandahong (Taiwan Bright Red)

Red-skinned with red flesh dragon fruit (Fig. 2.6). It is a self-pollinated Taiwan variety. The weight of a single fruit is 500 g with the maximum weight up to 1 500 g. The yield per mu is more than 4 000 kg, and the soluble solid content in the heart reaches 18%~22%. From the first flower buds in mid-April until the last batch of flowers withering in early December, there are about 16 batches of flowers in total. Its fruit can be left on the tree for more than 15 days in summer without cracking, and can be left on the tree for 2 months in winter. The best regions for its cultivation are Guangxi, Guangdong, Fujian, Hainan, and Yunnan provinces.

Fig. 2. 6   Taiwandahong (Taiwan Bright Red)

6. Shihuoquan (Stone Fire Spring)

Red-skinned with red flesh dragon fruit (Fig. 2. 7). It is a self-pollinated Taiwan variety. The weight of a single fruit is 446 g, with the maximum weight up to 1 050 g. The four-year-old plant has an annual yield of more than 7. 8 kg, and the soluble solid content is 20%. Buds appear in early and mid-May, and the flowering period

Fig. 2. 7   Shihuoquan (Stone Fire Spring)

lasts from late May to mid-October. The fruit maturity period is from early July to late November. A single plant can yield 4~8 batches of fruit throughout the year, and the fruit growth and development period is 30 ~ 40 days. The best regions for its cultivation are Guangxi, Guangdong, Fujian, Hainan, and Yunnan provinces.

7. Guizihonglong (Prosperous Purple-Red Dragon)

Red-skinned with red flesh dragon fruit (Fig. 2. 8). It is a self-pollinated variety evolved from the Zihonglong (Purple-Red Dragon) bud. The weight of a single fruit is 215 g, the yield per mu is 870 kg, and the soluble solid content is 13. 8%. The buds appear in early May, and the flowering period lasts from mid-May to mid-October. There are 8 ~ 10 batches of

Fig. 2. 8   Guizihonglong (Prosperous Purple-Red Dragon)

flowers and fruit each year. The first batch of flower buds appears in early May, bloom in mid-to-late May, and the fruit matures in late June. The last batch of flower buds appear in late September, bloom in mid-October, and the fruit ripen from late November to early December. The best regions for its cultivation are areas of Guizhou Province where the average temperature in January is above 0 ℃.

8. Ruanzhidahong (Soft Branch Bright Red)

Red-skinned with red flesh dragon fruit (Fig. 2. 9). It is self-pollinated. The weight of a single fruit is 500~1 000 g, the yield per mu is 4 000~5 000 kg, and the sugar content can reach more than 20%. The buds appear in early May, and the flowering period lasts from mid-May to

Fig. 2. 9   Ruanzhidahong (Soft Branch Bright Red)

mid-October. There are 13~15 batches of fruit bearing throughout the year. The fruit is harvested from June to February of the following year. The best regions for its cultivation are Guangxi, Guangdong, Fujian, Hainan, and Yunnan provinces.

9. Mihong (Honey Red)

Red-fleshed dragon fruit with super large fruit (Fig. 2. 10).
Selected and bred in Taiwan
Province. Its fruit is oblong,
dark purple-red. The seeds are
like black sesame. It has a
good taste and its storage period
is long. The weight of a single
fruit is 650 g, and the maximum
weight can be up to 1 540 g. In
the full fruit stage, the yield

Fig. 2. 10   Mihong (Honey Red)

per plant is 6. 6 kg, and there is no variation between years. Spring
shoots start to appear in early March, and new shoots stop
germination after entering the fruiting period. It has multiple
flowering and fruiting cycles within a year, buds appear in early
May, flowering period lasts from mid-May to mid-October, and
fruit maturity period lasts from early July to mid-January of the
next year, a single plant can yield 6 ~ 12 batches of fruit
throughout the year, and the fruit growth and development period
is 25~40 days. The best regions for its cultivation are Zhangpu of
Fujian province and Hainan province.

10. Qianguo No. 1 (Fruit of Guizhou No. 1)

Purple-red flesh new variety with large fruit from bud
mutation of the Zihonglong (Purple-Red Dragon) (Fig. 2. 11). Its
fruit is oval, with purple-red flesh and black seeds. The weight of a
single fruit is 460 g, the maximum can be up to 786 g, and the
soluble solid content is 13. 6%. The fruit is well colored, not easy
to crack, and has a strong flavor. There are 8~10 flowering and
fruiting cycles within a year. The first batch will bud in early May,
bloom in mid-to-late May, and mature in late June. The last batch

will bud in late September, bloom in mid-October, and ripen from late November to early December. The best regions for its cultivation are areas in Guizhou province where the lowest temperature in January is above 0 ℃.

Fig. 2. 11   Qianguo No. 1 (Fruit of Guizhou No. 1)

11. Fuguihong (Rich Red)

Red-fleshed dragon fruit (Fig. 2. 12). As a self-compatibility new variety selected and bred in Taiwan province, its self-pollination rate is up to 100%. Its fruit is large, oval, with thin skin. The weight of a single fruit is 445. 6 g, and the maximum weight can be more than 1 000 g. The yield of 4-year-old plant is more than 7. 8 kg. The skin of mature fruit is bright and rosy red, and the flesh is purple-red which is soft, crisp, and juicy. The soluble solid content of the fruit is 21% . It has strong

Fig. 2. 12   Fuguihong (Rich Red)

adaptability and its capacity of production has no variation between the years. Spring shoots start to appear from late March to mid-May, and the new shoots stop germination after entering the fruiting period; buds appear in early and mid-May, the flowering period lasts from late May to mid-October, and the fruit maturity period is from early July to late November. The output is mainly concentrated in July—August. A single plant can yield 6 ～ 12 batches of fruit throughout the year, and the fruit growth and development period are 30 ～ 40 days. The best regions for its cultivation are Guangxi and Guangdong provinces.

12. Mixuanlong (Honey Dragon)

New variety from Taiwan province, which is from the same family as Taiwandahong (Taiwan Bright Red) (Fig. 2. 13). It is self-pollinated and its fruit is almost round. Summer fruit tends to be large, and the largest winter fruit can be up to 1. 8 kg. It is a high-quality variety with red flesh

Fig. 2. 13   Mixuanlong
(Honey Dragon)

and not easy to crack. Its fruit is fragrant, juicy, with few seeds and high sugar content. The flesh tastes fresh and crisp.

13. Daqiu No. 4

Red-fleshed dragon fruit which needs artificial pollination (Fig. 2. 14). It is cultivated by Guangzhou Daqiu Organic Agriculture Limited Company and Fruit Tree Research Institute of Guangdong Academy

Fig. 2. 14   Daqiu No. 4

of Agricultural Sciences. Its fruit is nearly spherical, with a single fruit weight of 316 g and the soluble solid content of 9%. A 5-year-old plant yields 2 452 kg per mu. The best regions for its cultivation are the central and southern areas in Guangdong Province.

14. Hongxiuqiu (Red Hydrangea)

Its fruit is round (Fig. 2. 15), with a single fruit weight of 540 g and maximum weight up to 1 350 g. The mature fruit is bright red. The skin is 0. 3 cm thick and easy to peel off. The flesh is red and delicate, with small black sesame-like seeds. It is juicy and has a soluble solid content of 24%. Seedlings are planted in March. After 10~12 months, it will bloom and bear fruit, 12 ~ 15 times within a year. The yield per mu is 2 947 kg

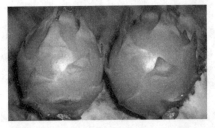

Fig. 2. 15   Hongxiuqiu (Red Hydrangea)

in the third year. Cultivation in greenhouse can yield all year round.

15. Meilong No. 2 (Beauty Dragon No. 2)

Bud mutation of the Hongcuilong (Red and Green Dragon) (Fig. 2. 16). It is self-pollinated with a fruiting rate up to 90%. Its fruit is nearly spherical, the skin is red with purple, thick and not easy to crack. Its flesh is dark purple-red, fine and extremely juicy, with a soluble solid content up to 20%. The weight of a single fruit is 500~ 1 000 g, and the yield per mu is 1 475 kg. It can yield 12~14

Fig. 2. 16   Meilong No. 2 (Beauty Dragon No. 2)

batches of fruit within a year, of which there are about 6 batches of large fruit. The first batch of fruit ripen in mid-June, the last batch of fruit ripen in early December. In summer, the fruit ripens 30~35 days after flowering. The mature fruit can be left on the tree for 10~30 days. Open field cultivation in Nanning of Guangxi is recommended.

16. Hongjinbao (Red Gold Treasure)

Self-pollinated, special variety selected by Taiwan (Fig. 2. 17). Because of its thick skin not easy to crack, it is easy to be stored and transported. Its unique scales make it look like flames of ruby. The weight of a single fruit is more than 1 300 g, and the maximum weight can be up to 2 200 g. Currently, it is the variety with the largest fruit among all red

Fig. 2. 17   Hongjinbao (Red Gold Treasure)

dragon fruit varieties in the world. The flesh has a subtle aroma, and the soluble solid content is 23%. The fruiting time is from June to December of the next year. If the late flowering period is prolonged by techniques, the fruit yield can be extended to February of the next year. The best regions for its cultivation are Taiwan, Nanning and Baise city of Guangxi.

17. Jirenmeigui (Lucky Rose)

It is bred by Nanning Dajili Agriculture Limited Company from high-quality Israeli dragon fruit and local high-quality wild dragon fruit (Fig. 2. 18). The flesh is crisp and sweet, with a rosy fragrance. The soluble solid content is more than 22%, and it is rich in anthocyanin. It is picked only after ripening on the tree. The scales of the fruit are very red, thin and wrinkled. After picking, no chemical treatment for preservation is required.

18. Guixianghong (Guangxi Fragrance Red)

Self-pollinated Taiwan variety (Fig. 2. 19). The weight of a single fruit is about 500 g with the soluble solid content up to 22%. Its yield per mu is 4 000 kg. With a fragrant odor, agreeable sweetness and delicate taste, it is not easy to crack and easy to be stored. 12~14 months after planting, it starts to bloom and bear fruit. It can flower 12~15 times each year. It takes about 30 days from flowering to fruit ripening. It can be bought in the market from June to December. Due to high disease resistance, it hardly requires pesticide, and that makes it can be an organic fruit. The best regions for its cultivation are Guangdong, Guangxi, Yunnan provinces, and most areas in Guizhou province.

Fig. 2. 18   Jirenmeigui (Lucky Rose)   Fig. 2. 19   Guixianghong (Guangxi Fragrance Red)

19. Xiangmilong (Fragrance Honey Dragon)

Self-pollinated with male and female styles of the same height. It has high fruit setting rate, and the biggest fruit can weigh up to

1 000 g (Fig. 2. 20). Its yield per mu is 5 000 kg. It is rich in anthocyanin, and has purplish flesh, fine texture and rosy fragrance. The soluble solid content can reach to 23%. Its flowering period starts from late May to early June and lasts to mid-November of the same

Fig. 2. 20　Xiangmilong (Fragrance Honey Dragon)

year, during which period of time, 8 to 13 batches of fruit can be harvested. It takes 30 ~ 40 days from the day of flowering to fruit ripening and ready for harvest. Because of its disease resistance, it can be planted in the open field in the south and in the greenhouse in the north.

20. Guihonglong No. 1 (Guangxi Red Dragon No. 1)

Red-fleshed dragon fruit (Fig. 2. 21) . Self-pollinated, it is selected and bred from the bud mutation of individual plant of the ordinary red-fleshed dragon fruit by Horticultural Research Institute of Guangxi Academy of Agricultural Sciences. Rosy skinned, dark purple-red fleshed, it gets tender texture.

Fig. 2. 21　Guihonglong No. 1 (Guangxi Red Dragon No. 1)

Juicy, clear and sweet, it has a bit rosy fragrance, with the soluble solid content is 22%. The weight of a single fruit is 533. 3 g. The yield per mu of the 4-year-old plant is 2 869. 75 kg. After ripening, it can be left on the tree for more than 15 days. Due to its strong adaptability, it can be selected and bred as cold resistant. Stable genetic character makes mass planting of single variety possible. The best regions for its cultivation are Guangdong, Guangxi and Hainan provinces.

21. Meilong No. 1 (Beauty Dragon No. 1)

Red-fleshed dragon fruit (Fig. 2. 22). The fruit bearing rate through natural pollination is 92%. Horticultural Research Institute of Guangxi Academy of Agricultural Sciences and Nanning Zhenqi Agricultural Technology Development Limited Company selected the quality individual plant from the seedlings of the hybrid combination of Costa Rica Red-Fleshed Dragon Fruit and Baiyulong (White Jade

Fig. 2. 22　Meilong No. 1 (Beauty Dragon No. 1)

Dragon) introduced from Vietnam. Its fruit is long and round, the skin is bright red, and the flesh is tender and fine. It tastes fresh, sweet, crunchy and vaguely fragrant. The fruit is medium-large size and the weight of a single fruit is 525 g. The yield per mu in the third year is 1 870 kg. The first batch of fruit ripens in mid-June, and the last batch ripens at the end of December. In summer, it ripens 30~35 days after flowering. The best regions for its cultivation are Guangxi and Guangdong provinces.

22. Guire No. 1 (Guangxi Heat No. 1)

Red-fleshed, self-pollinated dragon fruit (Fig. 2. 23). It is a variant selected from Gui-honglong No. 1 (Guangxi Heat No. 1) by Guangxi Subtropical Crops Research Institute and Guangxi Mountainous Area Comprehensive Technology

Fig. 2. 23　Guire No. 1 (Guangxi Heat No. 1)

Development Center. Its skin is bright red, its flesh is purple-red, and its bract has no spines. The weight of a single fruit is over 600 g, and the yield per mu in the third year is 2 015 kg. The best region for its cultivation is Guangxi province.

23. Chang'e No. 1 (Chang'e: Chinese Goddess of the Moon)

Red-fleshed dragon fruit with natural pollination (Fig. 2. 24). Its fruit is nearly oval, the skin is rosy, and the flesh is dark red in color and fine in texture. It is juicy, fresh and sweet. The weight of a single fruit is 410 g, and the yield per mu is up to 2 865 kg. The best region for its cultivation is Guangxi province.

Fig. 2. 24 Chang'e No. 1

24. Hongguan No. 1 (Red Crown No. 1)

Red-fleshed dragon fruit (Fig. 2. 25). It is bred via superior selection of individual plant from seedling propagation of Red Crystal Dragon Fruit introduced from Daqiuyuan Farm in Conghua District of Guangzhou by College of Horticulture of South China Agricultural University and Dongguan Institute of Forestry Science. Its fruit is almost spherical, and the weight of

Fig. 2. 25 Hongguan No. 1
(Red Crown No. 1)

a single fruit is 308 g. Its skin is bright red, and the flesh is purple-red, soft and smooth in texture. With the soluble solid content is 10.1%, it tastes fresh and sweet. The yield per mu in the third year is 1 628 kg. The best regions for its cultivation are the central and southern areas in Guangdong province.

25. Zihonglong (Purple-Red Dragon)

Purple-red fleshed dragon fruit (Fig. 2. 26). It is cultivated through systematic breeding by bud mutation of an individual plant of the Xinhonglong (New Red Dragon) by Guizhou Scientific Research Institute of Pomology. The fruit has round shape, purple-red flesh, and black seeds. The average weight of a single fruit is 330 g. The yield per mu in the third year is 2 000 kg. It

Fig. 2. 26　Zihonglong (Purple-Red Dragon)

can grow in four seasons and bear fruit 10~12 batches each year. It takes 15~21 days from budding to flowering, and 28~34 days from flowering to ripening. The best regions for its cultivation are the areas in Guizhou province where the annual average temperature is above 18.5 ℃ and the temperature in January is above −1 ℃.

26. Zilong (Purple Dragon)

Red-skinned with red flesh, self-pollinated dragon fruit (Fig. 2. 27). It is selected by Taiwanese growers through breeding among varieties that are introduced from Southeast Asia, Central America and other places and dozens of germplasm resources of red flesh dragon fruit collected locally. The weight of a single fruit is 245~850 g, and the yield per mu in the third year is 2 700 kg. The soluble solid content can reach 15%. The flowering and fruit bearing

period lasts from late March
to mid-November. In the phase
of fertility, there will be 12~
15 batches of flowering and
fruit bearing every year, and
the single cycle from flowering
to fruit bearing is 45 days. Its
flowering period obviously
overlaps the fruit bearing

Fig. 2. 27   Zilong (Purple Dragon)

period of time. For the same period, there can be up to 3 batches of
flowers and fruit. With high temperature, it ripens 35 days after its
flowering; with low temperature, it ripens 45 days after its
flowering. There are 4 major cycles of blooming within a year,
16~20 times to be more specific, 3~5 flowers on one branch at
one time. It takes 35~40 days for the growth and development of
the fruit. The best regions for its cultivation are Hainan, Guangxi,
Guangdong, Fujian and Yunnan provinces.

27. Jindu No. 1

Red-skinned with red
flesh, self-pollinated dragon
fruit (Fig. 2. 28). It is bred
and selected from the hybrid
variety of original seed of
Central American dragon fruit
and the red-fleshed dragon
fruit by Guangxi Nanning

Fig. 2. 28   Jingdu No. 1

Jinzhidu Agricultural Development Limited Company. It has short
and oval shape, purple-red skin, and dark purple-red flesh which
is tender and fine with a fresh and sweet taste and rosy fragrance. The
weight of a single fruit is 575 g, with the soluble solid content up

to 22%. The yield per mu is 3 820 kg. The buds come out in late April, and flowering period lasts from mid-May to late October. The ripening period lasts from late June to late December. For a single plant, there can be 8~12 batches for harvest throughout the year, and the period of growth and development of fruit lasts 35~50 days. The best regions for its cultivation are southern areas, southeastern areas and river valley areas in Baise of Guangxi province.

28. Tainong No. 3 (Taiwan Farmer No. 3)

Self-pollinated. Its fruit is round and its leafy scale is red (Fig. 2. 29). Thin skin, red flesh, it tastes smooth and tender. The

weight of a single fruit is 490 g. It has the best quality among the fresh-eating dragon fruit varieties. It can endure temperature as low as 0 ℃ and as high as 40 ℃ . The most favorable temperature for its growth is 25 ~ 35 ℃ . The best regions for its cultivation are Hainan,

Fig. 2. 29　Tainong No. 3 (Taiwan Farmer No. 3)

Guangxi, Guangdong, Fujian and Yunnan provinces.

29. Guanhuahong (Dongguan of China, Red)

Red-fleshed dragon fruit (Fig. 2. 30). It was bred via superior selection of individual plant from seedling propagation of Red Crystal Dragon Fruit by Dongguan Institute of Forestry Science and College of Horticulture of South China Agricultural University. Its shape

Fig. 2. 30　Guanhuahong (Dongguan of China, Red)

varies from nearly oval to spherical. The bright red skin is 0. 2 cm thick. The weight of a single fruit is 448 g. The flesh is purple-red and tastes smooth and soft with a strong fragrance. It is difficult to crack and easy to be stored and transported. The yield per mu for 3-year-old plant is 1 737 kg. The best regions for its cultivation are central and southern areas in Guangdong province.

Section Ⅱ   White Dragon Fruit

Ⅰ. Overview of White Dragon Fruit

Red-skinned with white flesh, self-pollinated dragon fruit. It has high resistance to diseases and a high fruit bearing rate. There is no spine on the skin of the fruit. The skin is thick, and the white flesh is fresh and sweet. It is easy to be stored and transported. Although the quality is average when it is consumed fresh, it can be processed into juice, fruit powder and jam.

White-fleshed dragon fruit includes: wild dragon fruit, Putongbaiyulong (Common White Jade Dragon), Xinbaiyulong (New White Jade Dragon), Hongbaoshi (Ruby), and Qianbai No. 1 (Guizhou White No. 1) etc. , among which Yuenan No. 1 (Vietnam No. 1) is the best. Its maturation period lasts from June to October. The best regions for its cultivation are Taiwan, Guangdong, and Guangxi provinces.

Ⅱ. Representative Varieties of White Dragon Fruit

1. Baiyulong (White Jade Dragon)

Red-skinned with white flesh dragon fruit which is a self-pollinated Taiwan variety (Fig. 2. 31). The skin is purple-red and 0. 2 cm thick. The flesh is white, and tastes clear, crunchy and juicy, with the soluble solid content up to 15%. The average weight of a single fruit is 425 g, with the maximum over 1 000 g. The full fruiting phase lies in the fourth year, with the yield per plant of more than 7 kg. From late March to late May, a large number of sprouts come out. In late May, the buds

Fig. 2.31   Baiyulong (White Jade Dragon)

appear, and it takes 13～18 days from budding to flowering. The flowering period lasts for over 4 months from mid-June to late September. The full blossom period is mainly from late June to early July, and from early August to late September. The harvest period lasts from late July to mid-November, and 6 batches of fruit can be harvested every year. The cycle of growth and development from flowering to ripening lasts 35～40 days. The best regions for its cultivation are Guangxi, Guangdong, Fujian, Hainan, Yunnan provinces.

2. Guanhuabai (Dongguan of China, White)

White-fleshed dragon fruit (Fig. 2.32). It was bred via superior selection of individual plant from seedling propagation of Red Crystal Dragon Fruit introduced from Daqiuyuan Farm in Conghua District of Guangzhou by Dongguan Institute of Forestry Science and College

Fig. 2.32   Guanhuabai (Dongguan of China, White)

of Horticulture of South China Agricultural University. It is neatly and nicely shaped. Its skin is light red and 0. 2 cm thick. The flesh is white, tastes agreeable and crunchy, fresh and sweet, with the soluble solid content up to 10. 8%. The average weight of a single fruit is 300 g, and the yield per mu in the fourth year is 2 405 kg. The best regions for its cultivation are the central and southern areas in Guangdong province.

3. Xianlongshuijing (Fairy Dragon Crystal)

White-fleshed dragon fruit (Fig. 2. 33). It was selected and bred from the Baishuijing (White Crystal) and Lianhuahong No. 1 (Lotus Flower Red No. 1) by Guangzhou Xianju Fruit Farm Agriculture Limited Company and Institute of Fruit Tree Research of Guangdong Academy of Agricultural Sciences. The skin is pink and 0. 3 cm thick. The flesh is white, tastes agreeable, fresh and sweet, with the soluble solid content

Fig. 2. 33   Xianlongshuijing (Fairy Dragon Crystal)

up to 11. 2%. The average weight of a single fruit is 325 g. The yield per mu in the fifth year is 2 933 kg. The best regions for its cultivation are the central and southern areas in Guangdong province.

4. Jingjinlong (Crystal Golden Dragon)

White-fleshed dragon fruit (Fig. 2. 34), which is also know as Qianbai No. 1. It is selected and bred via the bud mutation of individual plant of Jinghonglong (Crystal Red Dragon) in Luodian Dragon Fruit Farm by Guizhou Scientific Research Institute of Pomology. The flesh is white. There are red filaments in the parts close to the skin. It has fresh and fragrant flavor and tastes sweet. The average weight of a single fruit is 320 g, and the yield per mu in the third year is 1 685 kg. There are about 8 batches of fruit bearing every year, and for the same batch of flowering, its fruit maturity time is 5 days later than that of other varieties. The best regions for its cultivation are the areas in Guizhou province where the lowest temperature is above 0 ℃.

Fig. 2. 34　Jingjinlong (Crystal Golden Dragon)

5. Guanhuahongfen (Dongguan of China, Red Pink)

White-fleshed dragon fruit (Fig. 2. 35). It was bred via superior selection of individual plant from seedling propagation of Hongshuijing (Red Crystal Dragon Fruit) by Dongguan Institute of Forestry Science and College of Horticulture of South China Agricultural University.

Fig. 2. 35　Guanhuahongfen (Dongguan of China, Red Pink)

Its fruit is almost round, the skin is light red and the flesh is white. The part close to skin is pink. The soluble solid content is 11. 1%. The average weight of a single fruit is 239 g. The yield per mu in the third year is 1 169 kg. 25~45 days after the blossom fall, the fruit ripens. The best regions for its cultivation are the central and southern areas in Guangdong province.

6. Yuehong No. 3 (Guangdong Red No. 3)

Pinkish-white-fleshed dragon fruit (Fig. 2. 36). It was selected and bred from Baishuijing (White Crystal) and Lianhuahong No. 1

(Lotus Flower Red No. 1) by Institute of Fruit Tree Research of Guangdong Academy of Agricultural Sciences and Guangzhou Xianju Fruit Farm Agriculture Limited Company. 25 ~ 40 days after flowering, the fruit ripens. It is round, neatly and evenly shaped. The skin is pink and 0. 2 cm thick. The flesh is pinkish-white, and tastes tender, soft, fresh and sweet, with the soluble solid content up to

Fig. 2. 36   Yuehong No. 3 (Guangdong Red No. 3)

9. 54%. The average weight of single fruit is 285 g. The yield per mu in the fifth year is 2 530 kg. Spring sprouts come out from end of February to early May. The flowering period generally lasts from late May to late October. Four major batches of bloom are respectively in early June, early July, early September and early October. The ripening period starts from end of June to early December. Four major batches of ripening are in end of June, end of July, end of September and end of October. The best regions for its cultivation are the central and southern areas in Guangdong province.

7. Jinghonglong (Crystal Red Dragon)

White-fleshed dragon fruit (Fig. 2. 37). It is selected and bred via the bud mutation of individual plant of Putong-baiyulong (Common White Jade Dragon) by Guizhou Scientific Research Institute of Pomology. Its fruit is long oval shaped. The skin is purple-

Fig. 2. 37   Jinghonglong (Crystal Red Dragon)

red. The scales on the surface are yellow-green, flat and straight. The soluble solid content is 12. 0%. The average weight of a single fruit is 400 g. It grows in four seasons, and bears $7 \sim 9$ batches of fruit. From budding to flowering, it takes $16 \sim 18$ days; from flowering to ripening, $28 \sim 34$ days. The yield per mu is 1 452 kg. The best regions for its cultivation are Nanpan River, Beipan River and Red River Valley in Guizhou province.

8. Shuangse No. 1 (Double Color No. 1)

Red-fleshed self-fruitful dragon fruit with a white heart (Fig. 2. 38). It was bred via superior selection of individual plant from seedling propagation of Hongshuijing (Red Crystal Dragon Fruit) by College of Horticulture of South China Agricultural University and Dongguan Institute of Forestry Science. Its fruit is nearly spherical, and the skin is dark red. The central flesh is white and the rest is red. It tastes soft, smooth, crunchy, fresh and sweet with unique fragrance. The soluble solid content is 9. 8%. The average weight of a single fruit is $300 \sim 450$ g. The yield per mu in the third year is 1 850 kg. It is difficult to crack and easy to be stored and transported. The best regions for its cultivation are Guangdong and Hainan provinces.

Fig. 2. 38   Shuangse No. 1 (Double Color No. 1)

Section Ⅲ   Yellow Dragon Fruit

Ⅰ. Overview of Yellow Dragon Fruit

The yellow-skinned with white flesh variety is also known as Qilinguo (Kylin Fruit), Yanwoguo (Bird's Nest Fruit), and Huanglongguo (Yellow Dragon Fruit). It is of high quality when consumed fresh. Its flower period is long, and the flowers are large and fragrant. It takes 45~60 days from budding to flowering, 90~100 days from flowering to fruit maturity, and 110~150 days to fruit maturity in autumn and winter. The fruit will not crack, and it takes one month to turn from green to yellow (in winter). After turning yellow, it can be left on the tree for more than one month. It is the variety with the best quality, taste and sweetness. It has an aromatic taste, the seeds in the flesh are large and soft, and its soluble solid is more than 18%. However, its fruit is slightly smaller. The main fruit production period (autumn and winter fruit) is around the Spring Festival, and summer fruit are on the market around the Mid-Autumn Festival.

Yellow dragon fruit is divided into two categories: thorns and thornless. The yellow dragon fruit with thorns is further divided into Gelunbiyahuangqilin (Colombian Yellow Dragon Fruit) and Eguaduoeryanwoguo (Ecuadorian Yellow Dragon Fruit). The mature fruit of Colombian yellow dragon fruit is elongated, and that of Ecuadorian yellow dragon fruit is elliptical.

Ⅱ. Representative Varieties of Yellow Dragon Fruit

1. Eguaduoeryanwoguo (Ecuadorian Yellow Dragon Fruit)

Yellow-skinned with white flesh dragon fruit (Fig. 2. 39), which is also known as Golden Dragon. It is a self-pollinated variety with golden yellow color, and its flesh presents woven patterns resembling a bird's nest, which explains its Chinese name. This Ecuadorian variety has a flesh especially smooth and delicate. It has the best quality among all the dragon fruit varieties, with super sweet and aromatic taste. Its soluble solid content is more than 18%, and can reach

Fig. 2. 39　Eguaduoeryanwoguo (Ecuadorian Yellow Dragon Fruit)

25%. The winter fruit is on the market around the Spring Festival, and the summer fruit around the Mid-Autumn Festival. It takes 45~60 days from budding to blooming, 90~100 days from flowering to fruit maturity, and 110~150 days to fruit maturity in autumn and winter. It takes 120~200 days from flowering to fruit ripening. The weight of a single fruit is 350~450 g. The output is less than 1 000 kg per mu. It has strict requirements for planting conditions, and the cultivation cycle is as long as 15 months.

2. Gelunbiyaqilinguo (Colombian Yellow Dragon Fruit)

Yellow-skinned with white flesh dragon fruit (Fig. 2. 40), also called "Laohuanglongguo (Old Yellow Dragon Fruit)" in China, which needs artificial pollination. The unmature fruit is green, and the thorns will fall off when fully ripe. It is a rare variety. The flowering period

Fig. 2. 40　Gelunbiyaqilinguo (Colombian Yellow Dragon Fruit)

is from June to October every year, and the flowering and fruiting period are 160~210 days. The weight of a single fruit is 350~450 g. The yellow color and shape resemble kylin. The flesh is transparent and juicy, and can moisten the throat. The flesh presents filamentous pattern, smooth like a bird's nest. The soluble solid content is above 18%, and it has certain fragrance. Rich in dietary fiber and Vitamin C, it can eliminate oxygen free radicals. It also contains plant-based protein and anthocyanins, which are rare in other plants, and more iron than the other ordinary fruit.

3. Wucihuanglong (Thornless Yellow Dragon)

Yellow-skinned with white flesh dragon fruit (Fig. 2. 41), which is also known as Huanglongguo (Yellow Dragon Fruit), Huangmilong (Yellow Honey Dragon), and Dahuanglong (Big Yellow Dragon), it is called "Xinhuanglongguo (New Yellow Dragon Fruit)" in China. Self-pollinated, the flowering and fruiting rate can reach 90% on heavy rainy days. The fruiting period is from June to December each year. Introduced from Israel, golden-yellow colored, it is easy to be stored and

Fig. 2. 41   Wucihuanglong (Thornless Yellow Dragon)

transported. The fruit is large, not easy to crack with good taste. The flesh is crystal clear with a crisp and tender texture. Juicy, with strong fragrance, it tastes sweet with a little acid, sweet but fresh, which can improve digestion. The average weight of a single fruit is more than 250 g, the maximum can reach 500 g, and the yield per mu can reach 4 000 kg. It is currently a relatively drought-resistant and cold-resistant variety with a relatively high disease resistance in China. It is recommended to be planted in areas with high temperature differences between morning and evening at high altitudes. The best regions for its cultivation are Guangxi, Guangdong, Fujian, Hainan, and Yunnan provinces.

Chapter III
Traditional Cultivation Techniques of Dragon Fruit

Section Ⅰ Propagation Techniques of Dragon Fruit

Ⅰ. Seedling Cultivation from Cutting

1. Seedling Cultivation on Nursery Bed

Use a wooden box or directly use the land as the nursery bed. Fill in clean river sand of 15~20 cm deep, or adopt a mixture of 70% brick powder plus 30% charcoal powder as the substrate. The mixture of brick powder and charcoal powder shows a better performance, because the former has not only the drainage and air permeability as sand does, but also water absorption, a property that sand does not have, thus a certain degree of water retention; and the latter one has the function of making the propagation material anti-infective bacteria, as well as the properties of water permeability, air permeability and water retention. Therefore, the substrate made with the ration of these two materials is very effective for dragon fruit cutting. After the nursery bed is ready, cut a dragon fruit branch of 30 cm long and place it in a shady and cool place to dry naturally. The best time for planting is 7~10 days after the cut wound is cured. This can reduce the wound infection and rot, and the germination rate is up to 95%. Keep the soil dry and do not water it immediately after planting. Water it for the first time 10 days after planting.

2. Seedling Cultivation in Pots

Use a pot with a diameter of 17 cm. Pave its bottom with pumice or masonry of 3~7 cm deep, on top of which place a layer of newspaper with holes. Then spread a layer of snakewood sawdust or bark sawdust, or use about 3 cm-thick mixture of rice husk and soil with the ratio 3 ∶ 1. Put dragon fruit branch, and plant it on a slant; put wood chips or pine bark of 6. 6~10 cm on top, then put bricks. Water it after 10 days, and then water once every 3 ~ 4 days. It is suggested to water until water seeps out from the bottom of the pot.

Ⅱ. Seedling Cultivation from Grafting

1. Flat Grafting

Use a sharp knife to cut at the appropriate height of the triangular stem of Vileplume (Wild Trigonocarpus), and then cut the three peaks at 30°~40° degrees. Insert the disinfected cactus spine into the middle vascular bundle of the rootstock. The flattened scion is attached to the other end of the spine. Use the spine to connect the scion and the rootstock. The rootstock and scion should be as close as possible, leaving no gap in between, to avoid bacterial infection adverse to cure. Secure each side with spines and tie the base with thin wire (Fig. 3. 1).

Fig. 3. 1 Seedling Cultivation from Grafting

2. Wedge Grafting

Use a sterilized knife to cut a shallow slot at the top of the rootstock, and use a sterilized blade to cut the lower part of the scion into a duckbill shape and insert it into the slot of the rootstock. Fix it with plastic paper and put it in a plastic bag to keep the right air humidity, which is conducive to its survival. After 20 days of observing the growth after grafting, if it can keep fresh green, it will survive and can be out of the nursery in one month.

Ⅲ. Seedling Cultivation from Sowing Seed

Take a ripe dragon fruit, take out some flesh, soak it in water and crush the flesh. Wash it through the sieve several times and remove the flesh. Then filter it several times until the flesh is completely separated from the seeds. Sow the dried seeds on the soil and spray some water gently with a watering can. After 3~5 days, the seeds start to germinate and can be planted one month later.

Section Ⅱ   Establishment of Dragon Fruit Farms

Ⅰ. Choice of the Land

Choose a piece of land that is leeward, full of sunshine, fertile, with easy irrigation and drainage and elevation below 1 400 m . The plots rich in organic or improved sloped can be chosen as the farm (Fig. 3. 2) . According to the soil fertility, apply

Fig. 3. 2   Dragon Fruit Land

fully-decomposed organic fertilizer 2 000~3 000 kg per mu, and use tractor or tiller to evenly rotate to do the farming.

II . Preparation of Trellis

The dragon fruit is a vining plant. Trellis is needed to be built for the stem to climb. The production cycle of dragon fruit is usually more than 15 years, and the trellis material needs be strongly corrosion-resistant. Generally, cement columns of 10 cm × 10 cm are used as the trellis material, with a row spacing of 200 cm × 230 cm.

1. Cultivation with Single Column

Drill two holes with a diameter of 12~14 mm in shape of "+" where is 10 cm from the top of the cement column. The holes are left for the steel bars which are used like support rings. Each cement column needs to be equipped with 2 steel bars with a diameter of 12 mm or 14 mm and a length of 65 cm, and an old tire. To build columns at a density of 2 m × 3 m per mu, 111 columns are needed; at a density of 2.5 m × 2.5 m per mu, 106 columns are required.

2. Cultivation with Rows of Columns

With a wide spacing of (2.2 ~ 2.8) m × 3.5 m, 70 ~ 80 cement columns are needed for one mu. The cement columns need to be drilled crosswise at about 1m and 1.8 m height, and they will be further consolidated by steel wire. About 300 m of steel wire with a diameter of about 4 mm should be prepared for each mu of land.

III . Transplantation

1. Planting with Single Column

For planting with single column, the planting density usually is 2 m × 3 m or 2.5 m × 2.5 m, and the cement columns should be planted underground, with 1.6 m left above the ground. After the cement column is planted, fix it with soil, and apply 1. 52 kg of organic bacterial fertilizer containing 5% of N + P + K total nutrients, mix it with a hoe or a rotary tiller and pour it into a soil layer of 5~10 cm thick. Dig holes for planting around each cement column in four different directions. Each hole is 8 cm long and 6 cm deep and is for one dragon fruit plant. Leave the strong stems and buds of dragon fruit seedlings above the ground, plant its roots and cover them with soil, tie the seedlings to cement columns with more durable nylon ropes or cloth strips, and water them for rooting. With this cultivation method, for one mu, 420 ~ 450 plants can be planted and there are 5 000~6 000 fruiting branches.

2. Planting with Rows of Columns

Planting with rows of columns is usually based on row spacing of 2. 8 m×3. 5 m or 2. 2 m×3. 5 m. The column should be planted underground, with 1. 6 m left above the ground. The distance between each row of columns is 3. 5 m, and the surface of the columns perforated at 1. 8 m should face the same direction; after the columns are erected, the steel cable passing through the 1. 8 m holes between the columns connects the same row of cement columns into a main body, and the dragon fruit seedlings are planted, with a plant spacing of 0. 3~0. 4 m, between two columns on the planting row, supported by bamboo strips. When the plants are growing on the cables, the branches are pressed against the cables and parallel to each row. 10~12 branches per plant are kept. Use a 60 cm-long steel bar with a diameter of 12 mm or 14 mm to pass through the 1 m-opening, and use steel wire at both ends to straighten the steel bars between the same row of cement columns in the same direction. With this planting method, for one mu, 600~800 plants can be planted, and the fruit branches in high-yield area can reach a yield of 6 500~8 500 kg per mu.

Section Ⅲ　Dragon Fruit Sapling Management

After planting and pruning dragon fruit seedlings, replant for the missing ones and water in time when the soil is dry. After the dragon fruit survives, many new buds germinate on the branches. At this time, there should be only one strong bud left on each plant and the excess buds should be cut off in time.

Fig. 3. 3　Single Colum Planting

Ⅰ. Planting

The dragon fruit can be cultivated by single column or in rows.

If to cultivate by single column method, tie the newly grown stems to the cement columns with strings (Fig. 3. 3). When the stem grows to the supporting ring, you can cut off the heart to boost more branches to sprout. Leave $12 \sim 15$ branches on each plant to evenly distribute them on the supporting ring.

If to cultivate in rows, tie the newly grown stems to the cement columns with strings (Fig. 3. 4). When the plants grow on the pole, timely topping can boost the growth of new branches. Leave $10 \sim 12$ branches to distribute the

Fig. 3. 4　Row Planting

branches evenly to both sides of the supporting trellis, and remove the remaining buds.

Ⅱ. Fertilizer and Water Management

The dragon fruit grows rapidly in a warm and humid environment with sufficient sunlight. During the seedling growth period, the soil of the whole garden should be kept moist. Water more in spring and summer to keep the root system growth vigorously. During the fruit expansion period, the soil should be kept moist to facilitate fruit growth. Do not irrigate long time when watering, and do not pour water frequently during the growth period. Soaking will cause the root system to die after being hypoxic for a long time. In addition, drenching will cause uneven humidity and lead to erythema (physiological lesions). It is important to drain in time on rainy days to avoid the infection of germs which will cause the stems to rot. To enhance the cold resistance of the branches, the water should be limited in the orchard in winter.

Like other cactus, the growth rate of the dragon fruit is lower than conventional fruit trees. Therefore, the fertilization should be sufficient and be applied in small amount and multiple times. Nitrogen fertilizer is mainly suitable to 1—2-year-old saplings, and it should be applied in small amount and frequently to promote tree growth. Phosphorus and potassium fertilizers are suitable to mature trees over 3 years old, and the amount of nitrogen fertilizers should be controlled. Fertilization should be carried out during the shoot germination period in spring and fruit expansion period. The fertilizer is generally made of residue of oil crops left from oil extraction, chicken manure, and pig manure according to the formula of 1 : 2 : 7, and 25 kg of organic fertilizer should be applied to each plant every year. Or in July, October of the year and March of the following year, apply 1.2 kg of cow manure compost and 200 g of compound fertilizer to each plant.

The root system of the dragon fruit mainly grows at the surface soil layer, so the fertilizer should be casted, and deep furrowing and deep application should be avoided to prevent root damage. In addition, after each batch of young fruit is formed, 0.3% magnesium sulfate, 0.2% borax and 0.3% potassium dihydrogen phosphate are sprayed once above the roots to improve fruit quality.

Due to the long fruiting period of the dragon fruit, organic fertilizers should be applied in large amount. Nitrogen, phosphorus, and potassium compound fertilizers should be applied in a balanced proportion for a long time. If you only apply fertilizers rich in nitrogen like pig and chicken manure, the branches become thicker and turn dark green, which are brittle and easy to break in strong winds. Meanwhile, the fruit is large and heavy with poor quality and low sweetness. Some of them even taste sour or savory. Therefore, potassium fertilizer, magnesium fertilizer and bone meal should be applied during the period of flowering and fruiting to promote the accumulation of fruit sugar and improve the quality.

The dragon fruit has many aerial roots which can be turned into absorptive roots. Expanding the holes and improving the soil can gradually help expand the distribution of the root system. Moreover, we can tie the aerial roots and pull them to grow in the soil.

For young trees aged 1~2 years, mainly apply bio-organic fertilizers or composted cow manure and nitrogen fertilizers in small amount and frequently to promote tree growth. Apply 5~8 kg of mixed fertilizer containing high nitrogen, medium phosphorus and low potassium per mu, and water in time after application; or apply about 3 kg of water-soluble fertilizer containing high nitrogen, medium phosphorus and low potassium per mu. Apply fertilizer every 1~2 months together with watering according to the growth of the saplings.

Ⅲ. Key Points of Cultivation

There are two planting methods of dragon fruit: wall-support planting and shed planting. The column cultivation is the most common method, which is to plant 3~4 dragon fruit around the cement column, allowing the plants to grow upward along the column, and 110 columns can be erected per mu (the distance between row and column are 2.5 m × 2.5 m). Calculated by 4 seedlings per column, more than 400 seedlings can be planted in 667 square meters, which greatly improves the land utilization rate.

(1) Strengthen water and fertilizer management. The dragon fruit can be grown all year round. As its root system must be grown in well-aerated environment, it should not be planted deeply. Before planting, sufficient base fertilizer should be applied, and 10 kg of decomposed organic fertilizer plus 1 kg of compound fertilizer should be applied to each hole. The soil should be kept moist at the beginning, otherwise it is not conducive to growth. After planting, the soil should be trampled until it becomes firm, and the soil should be completely watered to facilitate rooting. Fertilizer should be applied in small amount at multiple times, and potassium and magnesium fertilizers should be supplemented during the flowering and fruiting period. When the temperature is low in winter, it is necessary to irrigate to keep the soil moist to promote rapid growth. Mainly apply organic fertilizer. Apply a small amount of chemical fertilizer, and fertilize 3 times in March,

July, and October respectively every year. Apply 2 kg of composted cow manure and 0. 2 kg of compound fertilizer to each plant to promote the accumulation of fruit sugar and improve the quality.

(2) Top and prune in time. When the branches grow to 1. 3~ 1. 44 m long, they should be topped to promote branching. Let the branches droop naturally for accumulating nutrients, so they can bloom and bear fruit early. After the fruit is harvested every year, the fruit-bearing branches are cut off to promote the growth of new branches to ensure the output in the next year.

(3) Prevent and control pests and diseases. The sapling of dragon fruit is susceptible to snails and ants. Insecticide can be used to control the pest infestation. In high-temperature and high-humidity seasons, it can easily be infected by diseases, resulting in partial necrosis of branch tissues and mildew spots. Fungicides can be used to prevent diseases.

Section Ⅳ　Shaping and Pruning of Dragon Fruit

The dragon fruit is a perennial tropical fruit, it is characterized by rapid growth, strong budding and branching ability, and long reproductive growth period. During the whole growth period, the conflict between vegetative growth and reproductive growth is particularly obvious.

As a very important dragon fruit cultivation and management technique, shaping and pruning are involved in the whole growth period. The shaping and pruning of dragon fruit are not a quick work that can be done once and for all, but a regular measure that should be adopted year by year to reasonably adjust the branch distribution and switch vegetative branches and fruiting branches according to the growth and fruiting of the plants. Proper and timely shaping and pruning are the basis for high quality and high yield.

Ⅰ. Overview of Shaping and Pruning

The shaping and pruning of dragon fruit are closely related and interdependent. The shaping is achieved through pruning, and pruning is done based on the foundation of a good shaping.

Shaping refers to the construction of the special structure and shape of branch based on the growth and development characteristics of dragon fruit with certain technical measures such as pruning. Pruning refers to the partially thinning out and cutting off of the stems, branches, buds, flowers and fruit of dragon fruit.

Ⅱ. The Effect of Shaping and Pruning

1. Enhance Ventilation and Light Transmission

As a plant which strongly favors sunlight, strong light is conducive to the flowering and fruiting of dragon fruit, and its outer branches would easily bear fruit. However, the dragon fruit grows vigorously and has strong sprouting ability. It is easy to grow a large layered canopy and grow outwards quickly, which will result in a serious overlap of branches, and insufficient light for inner branches, which is neither conducive to vegetative growth nor to reproductive growth.

By shaping and pruning, we can keep a proper number of branches and a suitable shape of the canopy, which will improve the lighting conditions, increase the coefficient of leaf space for photosynthesis, and promote the formation of strong fruiting mother branches, laying a good foundation for high yield.

2. Changes of Vegetative Branches and Fruiting Branches

The areoles on the branches of dragon fruit are mixed buds,

which can germinate branch buds and flower buds. In production, to increase the weight and improve the quality of single fruit and ensure sufficient nutrient supply for fruit development, the branches are artificially divided into fruiting branches and vegetative branches.

Select thick, plump, droopy branches of suitable-length as fruiting branches, which accounts for around 2/3 of the total number of branches; the rest should be vegetative branches, which accounts for 1/3 of the total number of branches; the newly germinated branches of the year can be cultivated as vegetative branches, which can be turned into fruiting branches in the next year.

The vegetative branches and fruiting branches are chosen according to the production needs and the growth state of the branches. When all the germinated flower buds on the fruiting branches are removed, the branches will be turned into vegetative branches. When the flower buds on the vegetative branches are reserved, the branches will be turned into fruiting branches. Through this way, the yield and quality of the fruit can be guaranteed.

3. Balance Vegetative Growth and Reproductive Growth

The yield of dragon fruit is affected by the length of the fruit-bearing period, the number of fruit and the size of the fruit. The single fruit development period of dragon fruit is short, however, the reproductive growth period of the whole cactus is long. Because of this, on a plant or even a fruit-bearing branch, large and small fruits, red and green fruits, large and small flowers, buds at different stages often coexist, they all compete fiercely for the nutrition.

Therefore, balance vegetative growth and reproductive growth through methods such as pruning, flower thinning and fruit thinning, properly adjust fruit yield during the cultivation, form sufficient vegetative areas, and maintain proper and strong growth of the tree are the keys to ensure a high yield and high-quality fruit.

Generally, more than 18 branches by plant should be kept during the full fruit yield period, in which more than 12 fruiting branches should be kept to ensure a balanced and average fruit-bearing of dragon fruit. During this time, apply 800-time-diluted Jiameihongli (plant root revitalizing regulator) 2~3 times per mu and apply 1 000-time-diluted Jiameizhiwunaobaijin (plant root revitalizing regulator) to promote root growth and nutrient transportation, and promote flower and fruit development.

4. Reduce Disease and Pest Damage

On the one hand, pruning enables vigorous growth of dragon fruit, which enhances the plant's ability to resist natural disasters and reduces its chances of being inflicted by infection of diseases, insect and pests. On the other hand, pruning itself is to remove diseased branches, weak branches and broken branches, which is one of the fundamental ways for pest and disease control.

Ⅲ. Methods of Shaping and Pruning

1. The Purpose of Shaping and Pruning of Young Trees

Make sure to put it on the climbing column as soon as possible to form an effective canopy system (Fig. 3. 5). The main measure is to keep a strong upward-growing branch, which is conducive to concentrate nutrition and quick up-on-the-column. When the main branch grows to a predetermined height, topping is adopted to promote branching to form a three-dimensional structure with proper space for its future growth.

The dragon fruit can germinate 15 ~ 20 days after planting, with an average daily growth of more than 2 cm. During its growth, many buds will grow out of the areoles. In the early stage, only one main branch is left to grow up along the column, and all other sided branches are cut off. When it grows to the required height of 1. 5~1. 8 m and exceeds the height of supporting disc or horizontal pole by 30 cm, topping is adopted to promote the growth of side branches from the top. Generally, about 3 buds are left on each branch, and the branches are guided to grow naturally through the disc or the horizontal pole. When the new bud grows to about 1. 5 m, top again to promote the secondary branching.

Pull or tie the upper branches to gradually make them to hang down, so as to promote the early formation of tree crowns and spacious growth space. 2~3 days are required for gradually increasing the number of branches, and finally keep 15~20 branches for each plant. The number of canopy branches for each column should be kept around 50~60. When the number of branches reaches a proper number as designed, as the side branches grow, the branches that are too dense on the side should be cut off in time to avoid too much consumption of nutrients.

Fig. 3. 5   Shaping and Pruning of Young Dragon Fruit Trees

2. Pruning during Vegetative Growth

There are two peaks in the vegetative growth of dragon fruit, which are mainly manifested by the germination of a large number of lateral buds and the thickening of stem nodes. Some are spring branches that germinate before flowering and fruiting in spring and summer (April-May); others are autumn branches that germinate after the flowering and fruiting end in autumn and winter (October-November). The purpose of pruning is to maintain the dynamic balance of the total number of reserved branches, timely renew fruiting branches and vegetative branches and promote the growth of fruiting branches.

After the spring branches germinate, the plant will enter the fruiting period as the light and temperature conditions become better, so pruning of the spring branches can reduce the consumption of nutrients. Generally, if a large number of old branches are in good conditions and will be kept, and there is a proper rate between fruiting branches and vegetative branches, all newly germinated buds should be removed as soon as possible so that the branches can enter the flowering and fruiting period as soon as possible.

If more fruiting branches are to be reserved, side branches from the base at the disc of the old branch can be kept for cultivation, and all the other new branches should be removed. Top in time when the new branches grow to about 1. 5 m. It is better to keep only one side branch on each old branch. The total number of side branches should not be more than 1/3 of the old branches. These branches can be cultivated into vegetative branches for flowering and fruiting in summer, and then be turned into fruiting branches in the spring of the following year. The replacement of sick and old branches should be carried out during the pruning of spring branches.

In autumn and winter, a large number of new branches will germinate. On the one hand, the redundant side buds should be thinned out, and buds at the base should be left properly to cultivate side branches of which the total number should be less than 1/3 of the old branches, to avoid the waste of nutrients. Top in time when the new side branches grow up to 1. 5 m to facilitate nutrient accumulation, so they can be used as vegetative branches in the spring of the next year, and can be turned into fruiting branches in summer. On the other hand, the branches that have already bear many fruit in the current year are less likely to bloom in large numbers and in a concentrated period in the next year. After the fruiting in autumn and winter, all the old branches that have fruited should be cut off, and new large and strong branches should be cultivated at the base of the stem. With the growth and drooping of the side branches, evenly distribute them on the disc or horizontal pole of the climbing column to build new fruiting branches group to ensure the output of the next year.

3. Pruning during Flowering and Fruiting Period

In production, if column cultivation is adopted, generally 50~60 drooping branches can be reserved for each cement column to form the fruiting branch group, which is 2/3 of the branches are kept as fruiting branches, and 1/3 of the branches are kept as vegetative branches respectively.

The plant enters the reproductive growth period from May to November every year, and it will bloom and bear fruit in batches continuously. Simultaneously, new branches will sprout and grow from the areoles and consume nutrients. The conflict between vegetative growth and reproductive growth is extremely obvious (Fig. 3. 6). The newly germinated lateral buds on the vegetative

Fig. 3. 6   The Newly Germinated Lateral Buds of Dragon Fruit

branches and fruiting branches should be thinned out to reduce nutrient consumption and improve sunlight conditions, thereby ensuring the nutritional supply for fruit development. At the same time, all the flower buds on the vegetative branches should be thinned out, the growth angle of the branches should be reduced, vegetative growth should be promoted, so they can be cultivated as strong fruiting branches of the next fruiting cycle.

4. Thinning of Flowers and Fruit

The dragon fruit has a long flowering period and strong flowering ability. It can bloom from May to November. During the full flowering and fruiting period, as many as 30 flower buds can grow out of one branch. The redundant flower buds need to be removed within 8 days after they appear. On average, only $3 \sim 5$ flowers buds are kept on each branch in every flowering cycle. After pollination and fertilization, the withered flowers can be cut off by ring cutting method (carving around the flower to stop nutrients transport), and the stigma and sepals below the ovary can be reserved.

Fruit thinning should be adopted when the young fruit's transverse diameter is about 2 cm. In principle, $1 \sim 3$ well-grown young fruits free from damage and deformity that have bright green color and a certain growth potential should be kept for each fruit-bearing branch, so as to concentrate nutrients and facilitate their growth (Fig. 3. 7). The excessive

Fig. 3. 7   Young Fruits of Dragon Fruit

fruits, deformed fruits or diseased fruits should be removed in time. During this period, 4 ~ 5 kg of Jiameihailibao (sea urchin fertilizer) should be sprayed per mu, and 1 000-times-diluted Jiameizhiwunaobaijin (plant root revitalizing regulator) should be sprayed at the same time to provide sufficient nutrition for the rapid expansion of the fruit and the germination of next batch of flowers buds.

IV. Attentions

1. Choice of Pruning Position

The length, number, and drooping angle of branches are the basis for high yield. Pruning is to build a good spatial structure for branches' growth, including the determining of the main stem, the length of vegetative and fruiting branches, and the distribution of drooping branches.

According to observations, the length of fruiting branches is generally longer than 1.5 m. The middle and upper branches, the tops of the branches and the drooping branches are most likely to bear fruit, however, the middle and lower branches rarely bloom. The growth of the upper branches is usually better than that of the middle and lower branches, it is probably related to the top advantage. Therefore, when topping or cultivating new branches, it is necessary to choose proper positions and lengths, and guide the branches to droop. Random topping or cultivation should be avoided.

2. Ensure Orderly Growth of Branches

Vegetative growth lays a foundation for reproductive growth. It is mainly manifested through the thickening of stem nodes, the increase in the number of branches and their elongation. The quantity and quality of vegetative branches are not only related to yield and quality, but also related to the replacement of fruiting branches and stable yield. Hence, we should not allow all branches to bloom and bear fruit out of the blind pursuit of large number of fruit while ignoring the preparation of vegetative branches. This disordered branch cultivation violates the application of quantitative cultivation management technique.

3. Quality of Pruning

The pruning of the branches and the thinning of the buds should be both carried out under the sun on a sunny day, so that the wound can heal easily and avoid the infection of germs. The pruning knife should be sharp and the cut should be quick and clean to avoid damaging the branches. All utensils should be disinfected with alcohol or potassium permanganate before pruning. Pruning, bud thinning, flower thinning and fruit thinning should be done in time to avoid late pruning after excessive nutrient consumption.

Section Ⅴ   Artificial Pollination of Dragon Fruit

For some varieties of dragon fruit, especially the red-fleshed dragon fruit, the stamen and style are of the same length or the stamen is shorter than style, so that the fruit setting rate through self-pollination is low. In addition, flowering at night makes insect pollination difficult, so artificial pollination is required.

The larger the genetic differences of different varieties of dragon fruit are, the higher pollination rate will be. Therefore, some different varieties of dragon fruit can be properly interplanted, especially red-fleshed and white-fleshed varieties can be mixed.

Ⅰ. Time of Pollination

The time of pollination should be decided according to the

flowering rule of dragon fruit. The dragon fruit usually blooms at 18:00 in the evening, and the flowering time is not long. Around 24:00, the color of the corolla begins to fade, the flowering generally lasts one night. Hence, the time for artificial pollination of dragon fruit should be carried out from the flowering in the evening to the morning before the flowers gradually close. It is better to pollinate at one o'clock in the morning, because at this time the diameter of dragon fruit flower is the largest, and the rate of pollination is also the highest.

Ⅱ. Methods of Pollination

The simplest method of pollination is to dip by Chinese brushes. Because the flowers of dragon fruit are relatively large, and each flower has much pollen, it is very simple to collect pollen. However, it should be noted that the viability of pollen is relatively low, and it will soon lose its viability in room temperature. As a consequence, during artificial pollination, it is advised to use the pollen once it is collected to ensure the viability of the pollen and pollination effect. After the pollen is collected, dip the pollen directly with a brush, and then spread it softly and evenly on the stigma.

Furthermore, it is a good method to put the container on the flower buds of dragon fruit first, tap the flowers, and the pollen will fall into the container. Then use Chinese brushes, brushes or other bristles to dip the collected

Fig. 3.8   Pollination

pollen and apply it to the pistil. Generally, the more pollen applied on the stigma, the larger the fruit of dragon fruit will be (Fig. 3. 8).

Cut off the petals 1~2 days after to avoid rot caused by thick petals and humid environment. When it rains during the flowering period, cover the buds and also cover them after pollination.

Section Ⅵ   Diseases Prevention and Control of Dragon Fruit

The dragon fruit has fleshy stems, large flowers, and rich sugar content, which provide conducive conditions for the reproduction of various microorganisms. Moreover, the dragon fruit grows in a hot and humid environment, which aggravates the problem of diseases.

Ⅰ. Stem Blight of Dragon Fruit

It is the main disease infecting the dragon fruit. It usually begins to develop in late March. During this period, the disease begins to appear, but the rate of diseased plants is still low and the disease develops slowly. From the beginning of April, the disease begins to aggravate. From mid-May to mid-July, it will peak. The disease favors high temperature and high humidity, and it is most severe from end of June to early July. It is much less active in winter after late November.

Ⅱ. Soft Rot of Dragon Fruit

It is most closely related to humidity. Rainy and humid areas or too much soil moisture will facilitate the reproduction, transmission and spread of germs, which will cause outbreaks of the disease.

In addition, temperature is a very important factor impacting the soft rot of dragon fruit. When the temperature is too low, the healing of wounds will be jeopardized, creating favorable conditions for bacterial infection. Generally, the onset begins in October, the peak period is from late January to early March, and the symptoms become less severe when the temperature rises in April of the second year. The disease is characterized by acute onset, rapid spread and great harm.

Ⅲ. Stem Spot Disease of Dragon Fruit

It mainly harms the fleshy stem of dragon fruit. High temperature and humidity create favorable condition for the disease, and excessive humidity is the main cause of the disease. Secondly, dragon fruit orchards with too dense branches, poor ventilation and light transmission, over-application of nitrogen fertilizers, and poor fertilizer management are more severely affected.

Ⅳ. Prevention and Control Techniques

The main prevention and control of dragon fruit's diseases should be comprehensive mainly based on agricultural control and supplemented by chemical control.

Agricultural prevention and control. The main methods are to establish a disease-free seedling nursery, select disease-free seedlings, and protect disease-free areas; plant and select new disease-resistant varieties; remove and destroy diseased residues collectively, remove weeds, reduce field humidity to reduce field pathogens; strengthen fertilizer and water management, increase the application of phosphate and potassium fertilizer to improve plant disease resistance. Among the three main diseases of dragon fruit, the soft rot is a bacterial disease, it has a fast onset and is not easy to control. In consequence, agricultural control is often used to prevent planting diseased seedlings, remove diseased plants in time, and therefore reduce the reproduction, propagation and spread of germs.

Chemical control. In the early stage of the disease, trim and scrape off the rotten parts in time, and then spray fungicides on the wound. The dragon fruit stem blight should be treated at the initial stage of the disease with one of the following pesticides at one of the concentrations: 500~800-time diluted 50% thiram WP or 50% methyl thiomycin. Spray once every 7 days, and spray 3 times in total. This method is proved efficient on the prevention and treatment of dragon fruit stem blight.

Section Ⅶ   Pest Prevention and Control of Dragon Fruit

The pests that harm the dragon fruit are mainly insects with piercing-sucking mouthparts and spiders (mites).

Ⅰ. *Nipaecoccus vastator* (Maskell)

This pest mainly infects new shoots, it concentrates on the edges of stems, and often lives on vines or branches where lack sufficient light or no light. It inserts the stem flesh with its mouthpart to absorb nutrients.

Control method: use a small bamboo stick with absorbent cotton or a small brush to remove the pests, apply concentrated eradication; wash with nicotine and soapy water; the method of spraying or sprinkling water can be adopted to prevent and limit the damages of *Nipaecoccus vastator* (Maskell). However, this method shall not be applied during the flowering period of dragon fruit.

II. *Aleurocanthus spiniferus* (Quaintanca)

Aleurocanthus spiniferus is Homoptera, Aleyrodidae. This pest mainly infects the stem. It sucks the juice from the stem and affects the growth of the plant. Once infected, white powdery adherents can often be seen on the tips and edges of dragon fruit stems, which are small white spots at first and which then gradually expand.

Control method: agricultural control measures include increasing organic fertilizers, combining nitrogen, phosphorus, and potassium fertilizers, timely pruning, continuously improving the fertilization, water, ventilation and lighting conditions of dragon fruit, to enhance the vitality of the plant and improve pest resistance. For the control of overwintering pest, apply 150 ~ 200-time-diluted 45% crystal lime sulfur mixture in early March and late October respectively to eradicate larvae and eggs and reduce the population base of pest in the field. During the peak period, 2 000-time-diluted 2.5% Bifenthrin solution or 3 000-time-diluted imidacloprid solution can be used. Spray twice at intervals of 10 days proves to be efficient.

III. Carmine Spider Mite

Carmine spider mite is Arachnoidea, Acarina, and Tetranychidae. Widely distributed and omnivore, it can damage more than 110 kinds of plants. In dragon fruit orchards, it mainly harms the new shoots of dragon fruit plants, which have filaments. The adult female mite is dark red, oval in shape, with black spots on both sides of the body. Overwintering eggs are red, and there are a small number of non-overwintering eggs which are pale yellow. The overwintering larvae are red, and the non-overwintering larvae are yellow. The overwintering nymphs are red, the non-overwintering nymphs are yellow, and there are black spots on both sides of the body.

Control method: once Carmine spider mites is found on dragon fruit plants, pesticides should be applied immediately. Pesticides such as profenate, pyridaben, etoxazole, bifenazate and abamectin can be used. Spray the medicine once every 5 to 7 days, and use it 2~3 times in a row. It is best to apply at dusk, spray it evenly on the whole plant. Switch the pesticide or use mixed pesticide can achive better effect.

Section Ⅷ   Harvesting of Dragon Fruit

30~40 days after the flowers have been successfully pollinated, the skin of the fruit begins to turn red and appears shiny. At this time, the dragon fruit can be harvested (Fig. 3. 9).

The dragon fruit should be harvested in time. Early harvest or late harvest will affect its quality. If it is harvested too early, the nutrients in the fruit have not been completely transformed, which will affect the quality and yield of the fruit; if too late, the fruit texture will become soft with less flavor and

Fig. 3. 9   Harvesting

declined quality, which will also affect the transportation and storage. A few weeks before the harvest, a harvest plan should be made, and all preparations for harvest should be carried out accordingly. The early-ripening fruit should be harvested first, and harvest in stages. The fruit for storage can be harvested earlier than the ones to be consumed fresh locally. The fruit for local sales and that will be processed can be harvested when they are fully ripe.

The dragon fruit should be harvested in favorable weather conditions, preferably after the dew dries up on a sunny morning with low temperature. If it is harvested on rainy or dewy days, there will be too much water on the fruit surface, which will cause diseases or pests infection. Harvest 2 ~ 3 days after days with strong wind and heavy rain. If it is harvested under the hot sun on a sunny day, the fruit temperature will be too high, which will cause strong respiration, and the quality of fruit during storage and transportation will be reduced.

The fruit scissors used for harvesting must have round tips to avoid stabbing the fruit. The fruit baskets should be lined with linen, paper, grass, etc. to minimize damage to the fruit.

When harvesting, use fruit scissors to cut off the fruit stem, and put the fruit gently in the basket or box (Fig. 3.10). The fruit stem should be kept. Fruit with stem lose less weight during storage than fruit without stem, and because their ripening process is

Fig. 3. 10　Storage

slower, the storage time is relatively longer. When the fruit stem is kept, we can cut off the fruit stem along the peduncle with fruit scissors, to avoid the stem to scratch other fruit during packaging, storage and transportation.

The harvested dragon fruit should be placed in a cool place, away from the sun and rain. After harvesting, the fruit should be generally classified and packed according to the size and fullness of the fruit. After being selected, classified and cleaned, the fruit should be put in carton or wooden crate. Put them one by one inside, fix the fruit and stack them in different layers, which can greatly reduce the damages of the fruit during storage and transportation, and can also improve the commodity grade of the fruit.

Generally, if the respiration rate of fruit is high, the storage time will be short, and vice versa. The harvested dragon fruit is a fruit with low respiration rate, and because of its thick and waxy skin, it can endure long storage and transportation. It can be stored at room temperature for more than 15 days. If it is packed in a box and refrigerated with storage temperature about 15 ℃, it can be stored for more than one month. However, in the hot season with high temperature, the dragon fruit must be placed in a shady place to cool and dissipate heat after harvesting, and be refrigerated to maintain fresh. Being easy to storage and transport makes the sale of dragon fruit more convenient and more competitive in the market.

Chapter IV
Modern Cultivation Techniques of Dragon Fruit

Modern cultivation techniques such as shed building, supplemental lighting and integrated irrigation system should be adopted to improve the yield and quality of the dragon fruit, expand the production area of dragon fruit, and increase the production efficiency and benefits of the dragon fruit.

Section I Protected Cultivation of Dragon Fruit

The dragon fruit is a tropical fruit. If introduced to the north, related protection facilities must adapt to create favorable conditions for its growth. Hence, it is necessary to choose a greenhouse with good heat preservation effect and sufficient lighting for planting. Commonly used cultivation facilities in northern cold regions include solar greenhouses with single-span and multi-span greenhouses, etc. In summer, shade cloth is used to reduce the interior temperature; and in winter, thickened cold-blocking curtains, insulation quilts or inner sheds can be installed, or heating facilities can be used to protect the fruits from extremely cold weather on snowy or windy days.

I. Facility Conditions

The multi-span greenhouse has a height of 2.8 m and a width

of 6. 5 m. The steel pipes of the greenhouse are all galvanized steel pipes with a diameter of 20 mm and a wall thickness of 2 mm. In summer, cover the pipes with single-layer white plastic film of 10 mm thickness. Add one layer of film inside in winter when the temperature is low. Double films can help maintain the temperature. Inner film is white plastic film of 4 mm

Fig. 4. 1   Protected Cultivation of
Dragon Fruit

thickness. The greenhouses generally have north-south orientation (Fig. 4. 1).

In the greenhouse, rein-forced cement columns with moisture resistance, high temperature resistance, corrosion resistance and strong support are used as dragon fruit climbing supports. The column specifications are 8 cm × 8 cm × 180 cm (length). The cement column is inserted deeply underground at a depth of 30 cm, 150 cm are kept above the ground, and the above-ground parts are fixed with 12 mm steel fibers, which can serve as climbing supports for dragon fruit's growth, flowering and fruit-bearing.

Install an automatic shutter machine, micro-sprinkler irrigation, indoor lighting and other facilities. Leave a drip irrigation hole next to each plant to realize the integrated supply of fertilizer and water and ensure timely adjustment and supply of water and fertilizer. To make full use of the space in the greenhouse, in the process of fertilization and furrowing, 3 ridges with a height of 20~25 cm are made for each shed, and a 30 cm-width ditch is left between the two ridges for agricultural field operation and irrigation.

II. Main Techniques

1. Moisture Management

Generally, the moisture content of soil during the growth period is 70%～80%, and the relative air humidity during the flowering period is 60%～70%. In the initial stage of planting, water every 2～3 days, and keep the soil moist during the seedling stage and fruit expansion stage to facilitate fruit growth. During drought, irrigate once every 3～4 days. Drain in time on continuous rainy days, so as to avoid infection which might cause the stem and the flesh to rot.

2. Temperature Management

When the temperature is between 17～25 ℃, it creates a favorable environment for the growth of dragon fruit. When the temperature difference between day and night is more than 10 ℃, it is conducive to flower bud formation and smooth pollination. When the temperature is between 25 ℃ and 38 ℃, it is conducive to fruit development. In order to ensure the year-round growth and multiple fruiting of dragon fruit, try to keep the temperature close to the above-mentioned values, maintain the temperature between 25 ℃ and 35 ℃, ensure that the night temperature in winter is not lower than 8 ℃, and pay attention to ventilation when it is hot in summer.

3. Light Management

The dragon fruit needs strong light, and generally requires light intensity of 8 000～12 000 lx. With insufficient light intensity, especially during the reproductive growth period, when the light intensity is below 8 000 lx, different degrees of supplemental lighting must be provided. However, if the light is too strong, moderate shading is required. To ensure all-year round flowering and fruiting of dragon fruit, supplemental light should be adopted.

Section Ⅱ   Supplemental Light Techniques of Dragon Fruit

The dragon fruit is a tropical fruit. In tropical regions such as Hainan province, it can bloom and bear fruit naturally from May to mid-December, but it cannot bloom and bear fruit naturally in winter from late December to April of the next year. To extend the fruiting period, supplemental lighting in winter can be adopted to increase production. The principle is to artificially imitate sunlight to promote the photosynthesis of dragon fruit and flower formation (Fig. 4. 2).

Fig. 4. 2   Supplemental Light Techniques

The sunlight threshold between flower bud initiation and differentiation of dragon fruit is usually close to 12 hours, and supplemental lighting is adopted when the daylight hours are shorter than 12 hours. After the autumnal equinox and before the spring equinox of the next year, the artificial light source is used to meet the demand of dragon fruit for daylight, to promote the dragon fruit to produce flower buds and increase the off-season yield.

Ⅰ. Selection of Light Supplement Lamp

When choosing the light supplement lamp for plants, LED lights are recommended, which are more energy-efficient (Fig. 4. 3). The supplemental light is generally yellow light combined with blue and red light. The wavelength of blue light is required to be around 430 nanometers to promote the formation of carotene and vitamins in fruit. The red light should be around 630 nanometers to promote flowering and fruiting. The combination of red and blue light is between 510 and 610 nanometers, with a color temperature between 3 000 and 4 000 Kelvin, 120 lx

Fig. 4. 3   LED Light Supplement Lamp

per square centimeter, and power of 12~18 W.

II. Position of Light Supplement Lamp

Choose the right height and distribute the lamps evenly. The height should be about 50 cm above the top of the plant. If it is installed too high, the lighting cannot reach the plants, and if too low, the upper branches will easily block the light. At present, the dragon fruit is generally cultivated in rows, with 1 000~1 500 plants per mu. The distance between the lights should be 1~1. 5 m width. The specific distance can be determined according to the power of the supplemental light, as long as the light intensity around the main fruiting branches is 120~150 lx.

III. Environmental Temperature

Temperature is a key factor influencing the flowering and fruiting of dragon fruit. When it is lower than 15 ℃, the dragon fruit can hardly bloom and bear fruit regardless of the supplemental lighting. When it is above 15 ℃, with supplemental lighting, the productivity can be boosted. The supplemental lighting is most effective when the temperature is above 20 ℃.

IV. Timing of Supplemental Lighting

To meet the demand of dragon fruit for light, currently the

lighting time in the field is more than 4 hours per day; after sunset between 18:00 and 22:00; darkness interruption 22:00~02:00; lighting 02:00~06:00. The actual number of lighting hours can be adjusted according to the variation of daylight time of the week. The lighting time can be adjusted between 2 and 4 hours accordingly, which can reduce lighting time and electricity costs.

In areas such as Yunnan and Guizhou provinces, the lighting time is generally divided by the autumnal equinox and the spring equinox. After the autumnal equinox, lamps are hung to supplement the light until the next spring equinox (Fig. 4. 4). In tropical areas like Hainan province, the temperature is relatively high, and the light is supplemented from mid-October to late March of the next year. It is suggested to

Fig. 4. 4   Supply Light after Sunset

provide supplemental lighting of 2~4 hours a day right after sunset. At this time, benefiting from the residual heat of the sun, about 2 hours of light supplement is enough. The later the lighting is provided, the longer the lighting should be lasted.

Section Ⅲ   Integrated Techology of Water and Fertilizer for Dragon Fruit

Agriculture is the pillar for economic development. In the future, water saving will still be the main direction of agricultural development. China will continue to increase its investment in promoting the development of water-saving irrigation technology. The promotion and application of integrated water and fertilizer irrigation will effectively improve China's agricultural irrigation, speed up the pace of modernization of agricultural irrigation in China and contribute greatly to the development of green, ecological and high-quality agriculture (Fig. 4. 5).

For a complete integrated water and fertilizer system, to ensure that the entire pipeline will not be blocked and the irrigation system runs smoothly, in addition to the water and fertilizer integration machine as the core, the filter is an essential part. Commonly used filters in water and fertilizer

Fig. 4. 5   Integrated Irrigation of
Water and Fertilizer

integrated systems include centrifugal filters, sand filters, mesh filters, laminated filters, etc. In addition to filters, there are also a series of accessories such as solenoid valves, irrigation pipes, and irrigation nozzles, etc.

The intelligent IoT water and fertilizer all-in-one machine is the core control center to realize fixed amount irrigation at fixed time with scientific calculation. While meeting the specific water and fertilizer needs of crops, it can increase crop yield and improve crop quality; at the same time, it can also save energy. The application of integrated water and fertilizer irrigation can save 50% of water, 40% of fertilizer, and 90% of labor on average, which will facilitate the orchard management and improve work efficiency.

Drip irrigation allows the water to flow evenly into the soil near the plant root system. The flow of water is low, and the water permeates slowly, which can minimize evaporation loss, and the water utilization rate can reach more than 95%.

Compared with drip irrigation equipment and sprinkler irrigation equipment, atomization micro-spray equipment has some unique characteristics. It can be applied to some scenarios in which drip irrigation and sprinkler irrigation cannot be applied, and achieve the effect that neither of them can. It is a water-saving irrigation method with great potential (Fig. 4. 6). The characteristics of micro-spraying are small amount of water, large irrigation range, low average water consumption, low pressure of micro-spray irrigation system. Therefore, large pressurization equipment is not needed, which is of low energy consumption. The water and fertilizer solution in the fertilization tank is filtered through a mesh filter and then sprayed out through the micro-spray nozzle, fertilizing the roots and leaves of the crops at the same time, which improves the utilization rate of fertilizers. The atomization refraction micro-nozzle irrigation saves more than 70% of water and more than 30% of fertilizer compared with traditional ground irrigation, and saves about 50% of energy compared with irrigation with large and medium nozzles.

Fig. 4. 6　Micro-spray Equipment

REFERENCES

Chengzhi Sun, 2005. Techniques of Protected Cultivation of Dragon Fruit [J]. Guangdong Agricultural Science (2): 2.

Jun Fan, 2015. Key Techniques of Protected Cultivation of Dragon Fruit in Chengcheng of Shaanxi Province [J]. Practical Technology and Information of Fruit Trees (12): 2.

Yonghua Qin, et al. , 2020. Colored Illustrations of Dragon Fruit Cultivation for High-Quality and High-Yield [M]. Guangzhou: Guangdong Science and Technology Press.

Yongqiang Cai, 2017. Key Techniques of Dragon Fruit Cultivation [M]. Beijing: China Agricultural Press.

Youjie Liu, 2019. Excellent Dragon Fruit Varieties and Efficient Cultivation Techniques [M]. Beijing: China Agricultural Science and Technology Press.

Yuhua Ma, and Yongqiang Cai, 2017. High-Efficiency Cultivation Techniques of Dragon Fruit [M]. Guiyang: Guizhou Science and Technology Press.